CONCEPTUAL REVOLUTIONS

CONCEPTUAL REVOLUTIONS

Paul Thagard

PRINCETON UNIVERSITY PRESS PRINCETON, NEW JERSEY

Published by Princeton University Press, 41 William Street,
Princeton, New Jersey 08540
In the United Kingdom: Princeton University Press, Oxford

Library of Congress Cataloging-in-Publication Data

Thagard, Paul.
Conceptual revolutions / Paul Thagard.
p. cm.
Includes bibliographical references (p.) and index.
ISBN 0-691-08745-8 (alk. paper)
1. Science—Philosophy. 2. Science—History. 3. Concepts—History. 4. Change (Psychology) 5. Biology—Philosophy.
6. Cognition. I. Title.
Q175.T4793 1992
501—dc20 91-20210 CIP

This book has been composed in Adobe Times Roman

Princeton University Press books are printed
on acid-free paper, and meet the guidelines
for permanence and durability of the Committee
on Production Guidelines for Book Longevity
of the Council on Library Resources

Printed in the United States of America

3 5 7 9 10 8 6 4

For Daniel

WHO WILL BE ELEVEN IN THE YEAR 2000

Contents

Figures

Tables

Acknowledgments

I AM GRATEFUL to many people for their help in developing the ideas in this book. Here is only a partial list: William Bechtel, William Brewer, Susan Carey, Lindley Darden, Myrna Gopnik, Steve Hanson, Keith Holyoak, Gilbert Harman, Daniel Hausman, Philip Johnson-Laird, Trevor Levere, Daniel Kimberg, Pat Langley, Michael Mahoney, Peter Meyers, Robert McCauley, George Miller, Greg Nelson, Nancy Nersessian, Gregory Nowak, Paul O'Rorke, Michael Ranney, Brian Reiser, and Jeff Shrager.

For valuable comments on a previous draft, I am particularly grateful to William Brewer, Lindley Darden, Philip Johnson-Laird, Noretta Koertge, Rachel Laudan, Robert McCauley, Gregory Nowak, Michael Ranney, Robert Richardson, and two anonymous referees.

During the period in which this book was written, my research was supported by contracts from the Basic Research Office of the Army Research Institute for Behavioral and Social Sciences, and by a grant to the Princeton University Human Information Processing Group from the McDonnell Foundation. The Princeton University Cognitive Science Laboratory provided an excellent working environment.

I am grateful to my research assistant Gregory Nowak for his historical analyses of the geological, Copernican, and Newtonian revolutions. Chapter 7 is based on articles written with him, and Chapter 8 also benefited from his research. Greg Nelson wrote the graphics program used to display ECHO runs. Malcolm DeBevoise of Princeton University Press and Ron Twisdale provided valuable assistance.

To my wife, Ziva Kunda, I am indebted for the title, numerous suggestions, and much more.

This project was conceived largely as a whole, but as it developed various parts became articles, which have now been excerpted, revised, and integrated into the current text. I am grateful to the respective publishers for permission to include vestiges of previously published articles:

Nowak, G., and Thagard, P. (forthcoming). Copernicus, Ptolemy, and explanatory coherence. In R. Giere (Ed.), *Cognitive models of science, Minnesota Studies in the Philosophy of Science*, vol. 15. Minneapolis: University of Minnesota Press. (chs. 4, 8)

Thagard, P. (1989). Explanatory coherence, *Behavioral and Brain Sciences, 12*, 435–467. Cambridge University Press. (chs. 4, 6)

Thagard, P. (1989). Extending explanatory coherence (reply to commentators), *Behavioral and Brain Sciences, 12*, 490–502. Cambridge University Press. (ch. 4)

Thagard, P. (1989). Scientific cognition: Hot or cold? In S. Fuller, M. de Mey, T. Shinn, and S. Woolgar (Eds.), *The cognitive turn: sociological and psychological perspectives on science*. Dordrecht: Kluwer Academic Publishers, 71–82. (ch. 5)

Thagard, P. (1990). Concepts and conceptual change, *Synthese, 82*, 255–274. Kluwer Academic Publishers. (chs. 2, 3)

Thagard, P. (1990). The conceptual structure of the chemical revolution. *Philosophy of Science, 57*, 183–209. (chs. 1, 2, 3, 5)

Thagard, P. (1990). Modelling conceptual change. In J. Tiles, G. McKee, and G. Dean (Eds.), *Evolving knowledge in natural science and artificial intelligence*. London: Pitman Publishing, 201–218. (chs. 1, 3, 8)

Thagard, P. (1990). Philosophy and machine learning. *Canadian Journal of Philosophy, 20*, 261–276. (ch. 2)

Thagard, P. (forthcoming). Philosophical and computational models of explanation. *Philosophical Studies*. Kluwer Academic Publishers. (ch. 5)

Thagard, P., and Nowak, G. (1988). The explanatory coherence of continental drift. In A. Fine and J. Leplin (Eds.), *PSA 1988*, vol. 1. East Lansing, MI: Philosophy of Science Association, 118–126. (ch. 7)

Thagard, P., and Nowak, G. (1990). The conceptual structure of the geological revolution. In J. Shrager and P. Langley (Eds.), *Computational models of scientific discovery and theory formation*. San Mateo: Morgan Kaufmann, 27–72. (ch. 8)

A final note: In July 1992, I become Professor of Philosophy at the University of Waterloo in Waterloo, Ontario, Canada.

CONCEPTUAL REVOLUTIONS

CHAPTER 1

The Problem of Revolutionary Conceptual Change

SCIENTIFIC knowledge often grows slowly with gradual additions of new laws and concepts. But sometimes science undergoes dramatic conceptual changes when whole systems of concepts and laws are replaced by new ones. By analogy with political upheavals, such changes are called *scientific revolutions*.

Although many historians and philosophers of science have stressed the importance of scientific revolutions, there has been little detailed explanation of such changes. How do conceptual revolutions occur? How can a new conceptual system arise and replace an old one? What are these conceptual systems whose transformation is so fundamental to scientific development? Are scientific revolutions rational?

I shall propose answers to these questions from a viewpoint that is psychological and computational. In an earlier book, I advocated the use of techniques derived from artificial intelligence (AI) to understand the structure and growth of scientific knowledge; I called the enterprise *computational philosophy of science* (Thagard 1988). Here I shall show the relevance of ideas from the cognitive sciences to the most dramatic phenomena in the history of science: scientific revolutions. The theory of revolutionary conceptual change developed is germane to central issues in cognitive psychology and artificial intelligence, as well as to disputes in the philosophy of science.

1.1 THE IMPORTANCE OF CONCEPTUAL CHANGE

In the philosophy and history of science, the question of revolutionary conceptual change became important with the 1962 publication of Kuhn's *Structure of Scientific Revolutions*. For several previous decades, philosophy of science had been dominated by the logical empiricist approach. Exemplified by such philosophers as Carnap (1950) and Hempel (1965), the logical empiricists used the techniques of modern formal logic to investigate how scientific knowledge could be tied to sense experience. Like the views of Popper (1959), the logical empiricists emphasized the logical structure of science rather than its psychological and historical development.

Kuhn, along with other historically inclined philosophers such as Hanson

(1958) and Feyerabend (1965), charged the logical empiricists with historical irrelevance. Kuhn (1970, 1) wrote: "History, if viewed as a repository for more than anecdote or chronology, could produce a decisive transformation in the image of science by which we are now possessed." Kuhn's general view of scientific revolutions and his accounts of particular scientific episodes must be questioned, but his basic claim that the development of scientific knowledge includes revolutionary changes can be sustained.

Kuhn's claims about scientific revolutions have caused great consternation among philosophers of science because of their apparent implication of irrationality. Kuhn so stressed the dramatic and noncontinuous nature of scientific change that transitions in scientific theories or "paradigms" took on the appearance of cataclysmic, nonrational events. According to Kuhn's early statements, later moderated, a scientific revolution involves a complete change in standards and methods, so that rational evaluation of competing views using external standards appears impossible. He even said that when one theory or "paradigm" replaces another, scientists work in a different world.

Although Kuhn's emphasis on revolutionary change was an antidote to the simplistic models of the logical empiricists, a finer-grained theory of revolutionary change than Kuhn presented need not succumb to irrationalism. To develop such a theory, however, we need tools different from both the formal ones of the logical empiricists and the vague historical ones of Kuhn. Artificial intelligence offers the possibility of developing such tools for describing the structure of scientific knowledge and the processes that advance it. We can begin to characterize the structure of conceptual systems before, during, and after conceptual revolutions; and we can investigate the cognitive mechanisms by which conceptual changes occur.

The importance of the problem of conceptual change is not restricted to the history and philosophy of science. Conceptual change is of general psychological interest, since people other than scientists also experience it. Children acquire much knowledge through observation and education, and developmental psychologists have recently been arguing that children's acquisition of knowledge is not simply a matter of accretion of new facts. Rather, it involves an important restructuring of their conceptual systems. Carey (1985) has suggested that children undergo a fundamental restructuring of their biological ideas between the ages of 4 and 10, and she explicitly compares this restructuring to scientific revolutions. Vosniadou and Brewer (1987, 1990) have similar speculations about children's learning of astronomy. McCloskey (1983) describes the difficulties of children and some adults in appreciating Newtonian physics. Chapter 10 shows that conceptual change in children can be understood within the same computational framework that sheds light on scientific revolutions, although scientific revolutions involve conceptual change that is more radical than what occurs in the ordinary cognitive development of children. Human learning and scientific discovery are continuous processes, but scientific revolutions are rare events that involve more dramatic changes than are experienced in everyday cognition.

My approach both to the history of science and to developmental psychology is computational. For many researchers in cognitive science, thinking can not only be modeled computationally like the weather and wind resistance: thinking *is* a form of computation. To model mental structures and processes, programs are designed with data structures corresponding to the postulated mental structures and with procedures corresponding to the postulated mental processes. Running the program and doing psychological experiments provides a way of judging whether the model corresponds to psychological reality.

Artificial intelligence offers numerous tools for constructing these kinds of cognitive models. The AI subfield of *knowledge representation* is concerned with techniques of representing information in a computer for intelligent processing. AI work on problem solving and planning is highly relevant to the problem solving activities of scientists. *Machine learning* is the AI subfield concerned with how computational systems can improve their performance by acquiring and modifying their structures and procedures. Ideas from knowledge representation and machine learning will figure prominently in the theoretical developments in later chapters.

We cannot, however, simply take over existing ideas from these subfields of AI. Although much work in machine learning has been done on topics such as concept formation, we shall see that available techniques are not adequate to account for the origins of many important scientific concepts. Moreover, AI researchers have concentrated on cases of learning by accretion of knowledge, rather than on cases of revolutionary replacement of complexes of concepts. New computational ideas are required to account for this kind of replacement. Previous work on scientific discovery (e.g., Langley, Simon, Bradshaw, and Zytkow 1987) has neglected conceptual change. Thus the theory of conceptual change developed here is an extension of research in machine learning, not just an application. The problem of conceptual change is open for the field of artificial intelligence, as well as for philosophy and history of science.

1.2 ARE THERE SCIENTIFIC REVOLUTIONS?

Many critics of Kuhn have challenged whether the concept of *revolution* is appropriately applied to the development of science (Toulmin 1972). The concept of revolution has itself undergone interesting changes, from its original application concerning objects such as celestial bodies going round and round, to modern usages involving political, social, and scientific changes (Cohen 1985; Gilbert 1973). The old view that a revolution was fundamentally a return to a previous state has been abandoned, and instead the term "revolution" is applied primarily to cases in which major transformations have occurred. Cohen (1985) argues that there have indeed been scientific revolutions, but his account is purely historical, judging developments in the history of science as revolutionary if scientists and historians have so described them.

I shall count conceptual changes as revolutionary if they involve the replacement of a whole system of concepts and rules by a new system. The two key words here are "replacement" and "system." Merely adding a new set of ideas poses no special problems, and replacement of a single concept or rule should be a simple process. What is much harder to understand is how one system can be replaced by another.

If knowledge in science were neatly accumulative, fact piling on top of fact, there would be no need for a theory of revolutionary conceptual change. But there are episodes in the history of science that strongly suggest the importance of conceptual revolutions. As principal data for my theory of revolutionary conceptual change, I shall take seven historical cases that have most universally been dubbed revolutions:

Copernicus' sun-centered system of the planets, which replaced the earth-centered theory of Ptolemy.

Newtonian mechanics, which, in addition to synthesizing celestial and earthbound physics, replaced the cosmological views of Descartes.

Lavoisier's oxygen theory, which replaced the phlogiston theory of Stahl.

Darwin's theory of evolution by natural selection, which replaced the prevailing view of divine creation of species.

Einstein's theory of relativity, which replaced and absorbed Newtonian physics.

Quantum theory, which replaced and absorbed Newtonian physics.

The geological theory of plate tectonics that established the existence of continental drift.

Examination of these cases in the light of the new cognitive theory of conceptual change will display, from a cognitive perspective, what is revolutionary about them. I am eager not to adulterate the overused term "revolution"; the importance of conceptual revolutions is so great in part because they are so rare. Science does not make revolutionary leaps very frequently, but when it does the epistemic consequences are enormous.

1.3 THESES ON CONCEPTUAL REVOLUTIONS

To introduce the theory of conceptual revolutions developed in later chapters, I now advance six theses that sketch my major claims. First, it is necessary to characterize scientific revolutions from a cognitive perspective. The concept of revolution was originally applied to scientific developments by analogy with political and social developments. Political revolutions involve major transformations in political structures; in the American Revolution, for example, power was transferred from the British monarchs and their representatives to American citizens. Social revolutions involve major transformations in social structure, with some social classes wresting wealth and power from other social classes, for example during the Chinese revolution of 1949 (Skocpol 1979). Similarly, to understand scientific revolutions, we need to have an understanding of the kinds of structures undergoing transformation. Thesis 1 accordingly states:

1. Scientific revolutions involve major transformations in conceptual and propositional systems.

But how are concepts organized? Since the pioneering work of Quillian (1968), it has been common in AI systems to have much of the organization provided by *kind* or *is-a* hierarchies. For example, Tweety is a canary, which is a kind of bird, which is a kind of animal, which is a kind of thing. Similarly, psycholinguists have noticed the importance of kind-hierarchies for organizing the mental lexicon (Miller and Johnson-Laird 1976), and in addition have emphasized the role that *part-whole* hierarchies play. Part-hierarchies have different inferential properties from kind-hierarchies: because canaries are a kind of bird and birds have feathers, you can generally infer that canaries have feathers, but you cannot infer that beaks have feathers because beaks are parts of birds. Part-hierarchies have not received nearly as much attention in AI, although Darden and Rada (1988) show their importance in the development of the notion of a gene. Nevertheless, part-hierarchies are important for organizing concepts because they provide orderings such as: a toe is part of a foot, which is part of a leg, which is part of a body. I therefore conjecture:

2. Conceptual systems are primarily structured via kind-hierarchies and part-hierarchies.

Chapter 2 discusses the nature of concepts and the hierarchies that organize them. It also argues that there is more to conceptual change than mere revision of beliefs.

From theses 1 and 2 follows the conjecture that all scientific revolutions involve transformations in kind-relations and/or part-relations. We shall see in later chapters that this conjecture is true. A thorough analysis of the chemical revolution displays major changes in both kind-relations and part-relations (Chapter 3). Darwin not only proposed a major reorganization of kind-hierarchies by reclassifying humans from being a special kind of creature to being a kind of animal, he also transformed the meaning of *kind* by substituting a historical conception based on common descent for a notion of kind based on superficial similarity (Chapter 6). In the geological revolution, plate tectonics brought with it reorganizations of the kind-hierarchies and part-hierarchies involving continents and the seafloor (Chapter 7). Revolutions in physics discussed in Chapter 8 also display transformations in kind-hierarchies and part-hierarchies. One major ingredient in the revolution wrought by Copernicus is the reclassification of the earth as a kind of planet. Newton differentiated mass from weight, and reconceived gravity as a kind of centripetal force. Einstein's relativity theory brought with it a conceptual organization very different from that found in Newtonian mechanics, viewing mass and energy as manifestations of mass-energy. Moreover, the meaning of part-relations was changed dramatically by the substitution of an integrated space-time for commonsense concepts of space and time. Finally, quantum theory blurred the distinction between wave and particles, since light waves are quantized and particles have wavelengths. In his most recent writings, Kuhn (1987) has identified changes in taxonomic categories as characteristic of scientific revolutions.

How do new concepts and laws arise? Empirical laws are usually framed in the same vocabulary as the observational descriptions on which they are based or in terms directly derived from observational ones. In contrast, theories often invoke entities and processes that are unobservable or at least unobserved. Electrons, quarks, and mental processes are examples of theoretical entities postulated because of the explanatory power of the hypotheses that state their existence. The distinction between theoretical and observational concepts is not absolute, since better instruments can render a theoretical entity observable, as the electron microscope did for the gene reconceived as a sequence of DNA. Concepts referring to theoretical entities or processes cannot be derived from observation, so how can they arise? Theoretical concepts can be formed by *conceptual combination*, in which new concepts derive from parts of old ones (Thagard 1988). For example, the concept of sound waves, which are not observable, is the result of conjoining the concept of sound with the concept of a wave, both derived from observation. I therefore conjecture:

3. New theoretical concepts generally arise by mechanisms of conceptual combination.

This conjecture seems to fit well with additional cases from the history of science, such as natural selection and continental drift, but I have not conducted a sufficiently complete canvas of scientific theories to feel confident that it is true. Other mechanisms are probably necessary.

Theses 2 and 3 deal with the structure and origin of conceptual systems, while theses 4 and 5 address the same questions for propositional systems. Chapter 4 will describe a computational theory of explanatory coherence designed to account for revolutionary and nonrevolutionary cases of theory evaluation. On this account, the most important relations between propositions concern explanation and contradiction. On the basis of that theory, I conjecture:

4. Propositional systems are primarily structured via relations of explanatory coherence.

Where do new hypotheses come from? In the nineteenth century, the philosopher C. S. Peirce coined the term "abduction" to refer to the formation of explanatory hypotheses, and computational research on abduction is growing rapidly (Josephson, Chandrasekaran, Smith, and Tanner 1987; O'Rorke, Morris, and Schulenburg 1990; Peng and Reggia 1990; Pople 1977; Thagard 1988). The simplest case of abduction is one in which you want to explain why an object has some characteristic and you know that all objects with a particular property have that characteristic. Hence you conjecture that the object has the property in order to explain why it has the characteristic. I save further discussion of abduction for Chapter 3, but here merely conjecture:

5. New theoretical hypotheses generally arise by abduction.

Because theoretical hypotheses contain theoretical concepts, they cannot be formed by generalization from observations.

Finally, we must address the question of how scientific revolutions take place. Revolutions involve such major transformations in conceptual and propositional organization that there is no way in which simple techniques for operating on initial representations can bring about the same change gradually. Rather, a new theoretical and conceptual framework must be developed somewhat independently of the existing one. But now we face the difficult question: by what mechanism does the new framework replace the old? I conjecture:

6. The transition to new conceptual and propositional systems occurs because of the greater explanatory coherence of the new propositions that use the new concepts.

Later chapters will show how the theory of explanatory coherence developed in Chapter 4 applies to Lavoisier's argument for the superiority of his oxygen theory over the accepted phlogiston theory, Darwin's argument for evolution by natural selection, and to all the other cases of scientific revolutions.

1.4 OVERVIEW

The theory of conceptual change summarized in the six theses is laid out primarily in Chapters 2–4. Chapter 2 reviews philosophical, psychological, and computational accounts of what concepts are and puts forward a theory of concepts compatible with revolutionary conceptual change. Chapter 3 discusses the nature of conceptual development and considers computational mechanisms by which new concepts and links between existing concepts are formed. Chapter 4 presents a theory of explanatory coherence that shows how replacement of conceptual systems can occur and describes ECHO, a computational model that implements the theory of explanatory coherence. Chapter 5 deals with residual issues from Chapter 4 concerning the relations between successive theories, the question of scientific rationality, and the nature of explanation.

Since the primary concern of this book is to present a theory of revolutionary conceptual change, not to survey the history of scientific revolutions, my discussion of the major revolutions is not chronologically ordered; an appendix to this chapter provides a chronology. Chapters 3 and 4 take as their major illustrative example the eighteenth-century revolution in chemistry, a particularly informative case of conceptual change. The nineteenth-century Darwinian revolution is analyzed in Chapter 6, and the twentieth-century geological revolution appears in Chapter 7. Chapter 8 ranges from the fifteenth-century work of Copernicus through the twentieth-century revolutions produced by the relativity and quantum theories. Obviously, the briefer treatment of revolutions in physics reflects only my own interests, not their relative importance in the history of science. The goal of this book is to provide a new cognitive perspective on scientific revolutions, not to duplicate the excellent narratives that historians of science have provided.

Chapter 9 considers whether there have been revolutions in psychology. The so-called behaviorist and cognitive revolutions that occurred in the twentieth century turn out to be quite different from the major revolutions in physics, chemistry, biology, and geology, in large part because they did not involve substantial theory replacement. Finally, Chapter 10 reviews the conceptual changes that occurred in the various scientific revolutions, and compares them with the kinds of conceptual change that developmental psychologists have attributed to children. Anyone who has read through the first section of Chapter 5 should feel free to skip around, reading later chapters in any order.

The aim of this book is to provide a theory of conceptual change that unites philosophical, psychological, and computational approaches and that applies to all the major scientific revolutions. Omitted is discussion of the *social* context of scientific change. I do not for a moment deny that social factors are important in the development of scientific knowledge; the relation between them and cognitive factors is briefly discussed in section 5.2.2. I also omit discussion of nonscientific conceptual revolutions. Perhaps other writers will explore whether the mechanisms of conceptual change and theory evaluation that are here used to explain scientific developments also apply outside natural science.

1.5 SUMMARY

Conceptual revolutions occur when whole systems of concepts are replaced by new ones. Scientific revolutions include the development of Copernicus' theory of the solar system, Newtonian mechanics, Lavoisier's oxygen theory, Darwin's theory of evolution, Einstein's theory of relativity, quantum theory, and the geological theory of plate tectonics. A theory of conceptual revolutions should illuminate these cases by saying what concepts are, how they are organized into systems, and how conceptual systems are formed and replaced.

1.6 APPENDIX: CHRONOLOGY OF REVOLUTIONS

4th century B.C. Aristotle.

2nd century A.D. Ptolemy.

1543	Copernicus' *On the Revolutions.*
1609	Kepler's *Astronomia Nova* proposes elliptical orbits.
1628	Harvey's *On the Motion of the Heart and Blood.*
1632	Galileo's *Dialogue Concerning the Two Chief World Systems.*
1644	Descartes' *Principles of Philosophy.*
1687	Newton's *Principia.*
1723	Stahl's *Fundamenta Chymiae.*
1772	Lavoisier, age 29, conjectures that calxes contain air.
1777	Lavoisier proposes that "pure air" combines with metal to produce metals.
1783	Lavoisier attacks the phlogiston theory.
1831–1836	Darwin travels on voyage of the Beagle
1837	Darwin concludes that species evolved.
1838	Darwin, age 29, discovers natural selection.
1859	Darwin's *On the Origin of Species.*
1871	Darwin's *Descent of Man.*
1887	Michelson-Morley experiment.
1905	Einstein discovers special relativity. Einstein uses a quantum hypothesis to explain the photoelectric effect.

1913	Bohr model of the atom. Watson introduces behaviorism.
1915	Einstein develops general relativity. Wegener's *Origins of Continents and Oceans* proposes continental drift.
1919	Bending of light is confirmed.
1923	de Broglie introduces particle-wave duality.
1926	Schrödinger's wave mechanics.
1927	Heisenberg's uncertainty principle.
1956	Cognitivist works by Miller and by Newell and Simon.
1959	Harry Hess postulates seafloor spreading.
1965	J. Tuzo Wilson develops concept of transform fault.
1968	Jason Morgan develops mathematical theory of plate tectonics.

CHAPTER 2

Concepts and Conceptual Systems

THIS chapter develops an account of the nature and organization of concepts that provides the basis for a theory of revolutionary conceptual change. It begins with a survey of the varied theories of concepts that have been proposed in philosophy. The major burden of this chapter is to contest the view that there is little need to discuss conceptual change because the development of scientific knowledge can be fully understood in terms of *belief revision.* This view is widely taken for granted in philosophy and artificial intelligence. A psychological discussion of the functions and nature of concepts, along with an account of the organization of concepts into hierarchies, shows how a theory of conceptual change can involve much more than belief revision.

2.1 PHILOSOPHICAL THEORIES OF CONCEPTS

From Plato to current journals in psychology and AI, discussions of knowledge and mind frequently deal with concepts, schemas, and ideas. A critical survey of all the different accounts of concepts that have been offered in philosophy, psychology, and AI would take a volume in itself. Fortunately, the task here is more focused: to give an account of concepts that can support a theory of conceptual revolutions.

Writers in different fields and at different times have had various goals behind their accounts of what concepts are. For philosophers, the goals have often been metaphysical and epistemological: to describe the fundamental nature of reality and to show how knowledge of reality can be possible. Psychologists have proposed theories of concepts in the service of the empirical goal of accounting for observed aspects of human thinking. AI researchers develop accounts of conceptual structure to enable computers to perform tasks that require intelligence when done by humans. I share all these general goals, but in this project they are subordinated to the particular problem of making sense of revolutionary conceptual change.

In epistemology and science as in politics, to be ignorant of the past is to be condemned to repeat it. The problem of the nature of concepts has been with us since Plato and Aristotle, and while summarizing the views of major philosophers at a one-paragraph level can be misleading, I want to give a general overview of the range of opinions concerned with what concepts might be.

The first influential theory of concepts was proposed by Plato twenty-three hundred years ago and was explicitly designed to *avoid* problems of change. Heraclitus had argued that the world was in a state of perpetual flux, and Plato concluded that valid knowledge would have to be of a heavenly realm of ideas removed from the changing, subjectively perceived world. In dialogues such as the *Euthyphro*, *Meno*, and *Republic*, Socrates served as Plato's mouthpiece in investigating questions about the nature of piety, virtue, knowledge, and justice (Plato 1961). Typically, Socrates argued that understanding of these concepts is not to be had through particular examples or through rough definitions. Plato's view was that knowledge of conceptual essences could only come from the use of education to regain the acquaintance with the heavenly forms that was lost at birth. Learning is then just recollection of the essences of things, and concepts are abstract, unworldly, and immutable.

Aristotle's *Metaphysics* (1961) systematically criticized Plato's theory of eternal ideas, arguing that there was no need to postulate an eternal substance. Aristotle proposed instead that only particular objects were substances: there are no universal concepts independent of the objects, since universals are only what is common to all things to which a predicate applies. The essence of *human* is to be found by a definition that states what all humans have in common, not by acquaintance with a heavenly idea of the human. In Aristotle we find fundamental worries about the relation of the general and the particular that have survived to this day. Should a concept be treated as something additional to all the objects that fall under that concept? Related questions arise today in cognitive science, when researchers debate whether there needs to be a mental representation of a concept over and above the representation of the objects that fall under it.

For Plato and Aristotle, the point of discussing ideas was metaphysical, not psychological: the issue was whether ideas are a separate kind of substance, not whether they are part of thought. In contrast, seventeenth-century discussions of ideas tied them closely to thinking. In his *Meditations*, Descartes said that ideas are thoughts that are like images of things, as when one thinks of a man or an angel (Descartes 1980, 68). Some of these ideas are innate, while others derive from an external source and others are constructed by the thinker. Descartes used his method of doubting everything to determine what ideas are "clear and distinct" and therefore safe sources of truth. For Descartes as for Plato, the most important knowledge was true *a priori*, independent of experience. Leibniz also treated an idea as something in the mind (Leibniz 1951, 281) and maintained that knowledge is primarily found through clear innate ideas.

Writing in the seventeenth, eighteenth, and nineteenth centuries, the British empiricists offered a very different view of the origins of ideas and knowledge. Locke, Berkeley, Hume, Hartley, Reid, and Mill all contended that ideas derive from sense experience. Locke's seventeenth-century *Essay Con-*

cerning Human Understanding claimed against Descartes that there are no innate principles in the mind. He characterized an idea as "whatsoever is the object of the understanding when a man thinks" (Locke 1961, vol. 1, p. 9). Ideas are divided into simple and complex, with simple ideas coming directly from sensation and complex ideas being formed from the mind out of simple ones. Not all thinkers of this period, however, adopted a mentalistic view of concepts. According to Weitz (1988), the seventeenth-century thinker Hobbes understood concepts as linguistic entities given through definitions.

Like Locke, Berkeley maintained in 1710 that the objects of human knowledge are either ideas imprinted on the senses or ideas formed by help of memory and imagination (Berkeley 1962, 65). He differed from Locke primarily in denying that there are any abstract ideas separate from ideas of particular things. A few decades later Hume distinguished the perceptions of the human mind into impressions and ideas: the former are forceful and lively perceptions while the latter are faint images of perceptions (Hume 1888, 1). Hume approved of Berkeley's rejection of abstract ideas, and extended Berkeley's skeptical conclusion that there is no material substance independent of ideas with an argument that the self should only be considered to be a bundle of sense impressions. Hume made a sharp division of objects of human reason into *matters of fact* that are acquired directly through sense experience and *relations of ideas* that arise by thinking with ideas acquired by sense experience.

Partly in response to Hume, Kant gave an account of the contents of mind that is often viewed as a synthesis of the rationalist account of Descartes and the empiricist account of Locke. He thought that there are pure, a priori concepts of the understanding, but that these can be applicable to sensory appearances by means of *schemata*, which are obscurely characterized as a rule of synthesis or a universal procedure of the imagination (Kant 1965, 182). Like the empiricists, Kant was concerned with thought, but he placed more emphasis on the a priori preconditions of thought. Like all his predecessors, Kant had a static view of concepts, considering in the abstract the conditions of their application.

Consideration of conceptual dynamics began only in the nineteenth century; Hegel probably deserves to be understood as the founder of the study of conceptual change. Whereas Kant tried to find a foundation for knowledge using both empiricist and rationalist ideas, Hegel stressed the importance of conceptual development. From Plato to Kant, philosophers were looking for a method to achieve direct and certain knowledge, either by reason or from the senses. Hegel for the first time rejected the need for any direct foundation for knowledge, emphasizing instead the need for consciousness to develop a grasp of truth through a dialectical process of passing through and criticizing successive stages of complexity, from sense experience to much more abstract thought (Hegel 1967). Even less than the other philosophers I have been outlining, Hegel's view of concepts does not lend itself to short summary, but deserves emphasis here for its dynamic character. Hegel's concepts, however, do not inhabit individual minds, but are part of the general development of reality (see Taylor 1975 for a lucid discussion).

Although Hegel's epistemological and historical views influenced thinkers as diverse as Karl Marx and C. S. Peirce, his account of concepts seems to have had little effect. Far more influential was the approach of the nineteenth-century mathematician and logician Frege, who is generally considered the founder of modern analytic philosophy. More than anyone else, Frege is responsible for the abandonment (until recently) of psychologism in twentieth-century philosophy. We saw that Plato's and Aristotle's views of ideas were not psychological, whereas from Descartes to Kant and Hegel ideas and concepts were part of mind. Frege's main concern was the nature of mathematical truth, and he strongly objected to any whiff of psychologism. Of sensations and images he stated (Frege 1968, v–vi): "All these phases of consciousness are characteristically fluctuating and indefinite, in strong contrast to the definiteness and fixity of the concepts and objects of mathematics." He was convinced that psychologism leads to the abandonment of objective knowledge: "We suppose, it would seem, that concepts sprout in the individual mind like leaves on a tree, and we think to discover their nature by studying their birth; we seek to define them psychologically in terms of the nature of the human mind. But this account makes everything subjective, and if we follow it through to the end, does away with truth" (Frege 1968, vii).

Instead, Frege developed the logical account that a concept is a function (in the mathematical sense) whose value is always a truth value (Frege 1970, 30). For example, the concept *even* is a function that maps the object 2 to the truth value true and the object 3 to the truth value false. Frege also said that concepts such as *even* are the reference of predicates such as "even." Just as Plato sought the heavenly ideas to ground knowledge of a changing world, so Frege moved to fix mathematical knowledge in an entirely logical framework in which concepts had no taint of the psychological.

In the 1920s, Carnap (1967) attempted to put empirical knowledge on a sound foundation by producing a logical derivation of the concepts of all fields of knowledge from concepts that refer to what is immediately given in sensory experience. The logical apparatus developed by Frege, Russell, and Whitehead appeared powerful enough to show that most concepts are definable in terms of direct observation. Several decades of work by Carnap, Hempel (1952, 1965), and others showed, however, that the dream of rationally reconstructing scientific knowledge from experience is not realizable: we cannot connect concepts with experience and the world with strict definitions that provide necessary and sufficient conditions for applications of the concepts.

Wittgenstein, in his later writings published in the 1950s, moved away from the Frege-Russell program of finding precise definitions for concepts. Rejecting Aristotle's view that ideas have essences, Wittgenstein claimed there need not be anything common to all the instances of a concept. There are many different kinds of games, for example, and all we can expect to find among them are family resemblances, "a complicated network of similarities overlapping and criss-crossing" (Wittgenstein 1953, 32e). He remarked on "the fluctuation of scientific definitions: what today counts as an observed concomitant will tomorrow be used to define it" (37e–38e). Similarly, Quine (1963) rejected the distinction made by Hume and Kant between truths of definition and truths of fact. Wittgenstein thus had a more flexible view of concepts than is found in earlier writers, but his account is more linguistic than psychological, addressing the question of the use of terms more than the question of the contents of mind. This, of course, was compatible with the prevailing behaviorist psychology of the time.

For Plato and Frege the great advantage of metaphysical or logical approaches to the nature of concepts was the escape from the mutability of the mundane and psychological. For understanding scientific development, however, we obviously need an account of how concepts can change. This account should incorporate the flexibility in matters of meaning and fact urged by Wittgenstein and Quine, but go beyond these philosophers in psychological and computational directions. Once we abandon the Platonic pursuit of the chimera of metaphysical certainty, there is no reason not to situate human knowledge in the minds of human beings and to use the resources of empirical psychology to build an account of the structure and growth of knowledge. In the current state of the theoretical side of this endeavor, computational models are essential for developing detailed views of what the contents and processes of mind might be.

Since psychological and computational views about the nature of concepts will be examined later (2.4–2.6), there is no need to summarize them here. But it is useful to review the different positions on concepts that have been proposed by philosophers and see how psychologists have arrived at different as well as similar views. Figure 2.1 is a taxonomy of different views of the nature of concepts; it is based on a survey by Weitz (1988), although I have diverged somewhat from his categories and interpretations. At the top, we can distinguish views that treat concepts as entities from ones that do not. Plato, Frege, and Hegel all considered concepts to be nonnatural, supersensible entities beyond the ordinary world of space and time. Aristotle considered concepts to be abstracted entities, while Hobbes and perhaps Wittgenstein considered them to be linguistic entities. Many theorists have taken concepts to be mental entities, but they have differed in whether concepts are taken to be largely innate or learned, and in whether concepts are *closed*—definable in

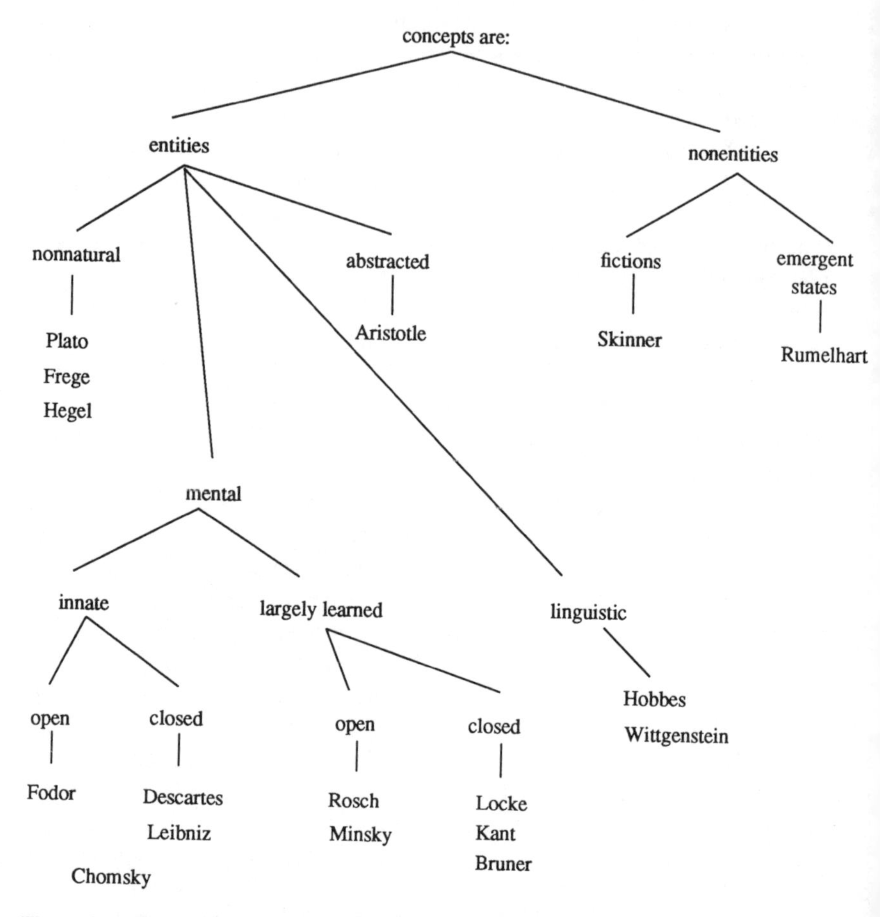

Figure 2.1. Taxonomy of theoretical views of the nature of concepts. Concepts are *closed* if they are definable in terms of necessary and sufficient conditions, and open otherwise.

terms of necessary and sufficient conditions. Among modern theorists, few besides Chomsky and Fodor hold that concepts are innate. Chomsky (1988, 134) claims that in addition to setting grammatical parameters "the language learner must discover the lexical items of the language and their properties. To a large extent this seems to be a problem of finding what labels are used for preexisting concepts." It is not clear whether Chomsky believes that concepts are open or closed. The innateness of concepts has also been defended by Fodor (1975), who maintains that concepts are open (Fodor, Garrett, Walker, and Parkes 1980).

Today, most theorists who view concepts as mental entities hold with Locke that concepts are largely learned from experience, while agreeing with

Kant that such learning presupposes an innate apparatus. Much psychological research has been conducted on concept learning, initially under the assumption that concepts are definable in terms of necessary and sufficient conditions (e.g., Bruner, Goodnow, and Austin 1956). In the 1970s, however, experiments by Rosch and her colleagues (Rosch et al. 1976) were used to support the view of concepts as *prototypes* lacking sharp defining conditions. Related critiques of the traditional view of concepts and definitions were mounted by Wittgenstein (1953), Putnam (1975), and Minsky (1975).

Theorists who deny that concepts are entities fall into two camps. The most radical camp claims that concepts are mere fictions and should not be part of a scientific analysis of language and mind. Skinner (1976), for example, rejected the postulation of mental entities such as concepts in favor of a focus on verbal behavior; see section 9.2 for further discussion of behaviorism. A novel nonentity view has emerged in recent years from connectionists such as Rumelhart and his colleagues, who assert about concept-like schemata: "Schemata are not 'things.' There is no representational object which is a schema. Rather schemata emerge at the moment they are needed from the interaction of large numbers of much simpler elements all working in concert with one another. Schemata are not explicit entities, but rather are implicit in our knowledge and are created by the very environment that they are trying to interpret—as it is interpreting them" (Rumelhart, Smolensky, McClelland, and Hinton 1986, 20). Concepts, then, are not entities but emergent states of neural networks. The connectionist view of concepts is discussed in section 2.4. My own view, developed later in this chapter, is that concepts are mental entities that are largely learned and open.

2.2 BELIEF REVISION VERSUS CONCEPTUAL CHANGE

Although discussions of concepts have been important in the history of philosophy, they have been much rarer in contemporary philosophy. According to Ian Hacking (1975), current analytic philosophy is the "heyday of sentences." Whereas seventeenth-century thinkers talked of *ideas*, contemporary philosophers take sentences to be the objects of epistemological investigation. Knowledge is something like true justified belief, so increases in knowledge are additions to what is believed. Epistemology, then, consists primarily of evaluating strategies for improving our stock of beliefs, construed as sentences or as attitudes toward sentence-like propositions. For example, Gärdenfors (1988) models the epistemic state of an individual as a consistent set of sentences that can change by expansion and contraction.

In the cognitive sciences, however, the intellectual terrain is very different.

In cognitive psychology, the question of the nature of concepts receives far more attention than the question of belief revision. Researchers in artificial intelligence often follow philosophers in discussing belief revision, but they also pay much attention to how knowledge can be organized in concept-like structures called *frames* (Minsky 1975; for reviews see Thagard 1984, 1988). Nevertheless, even a philosopher like Alvin Goldman (1986) who takes cognitive science very seriously places belief revision at the center of his epistemology, paying scant attention to the nature of concepts and the question of conceptual change. Gilbert Harman has written both on epistemic change (1986) and on the nature of concepts (1987), but has not much discussed the relevance of the latter topic to the former. Historically oriented philosophers of science such as Kuhn (1970) have suggested the importance of conceptual change but have not provided accounts of conceptual structure that are sufficiently developed for epistemological application.

The central question in current epistemology is: when are we justified in adding and deleting beliefs from the set of beliefs judged to be known? Without denigrating this question, I propose that epistemology should also address the question: what are concepts and how do they change? Concepts are relevant to epistemology if the question of conceptual change is not identical to the question of belief revision. But maybe it is; consider the following argument.

> The issue of conceptual change is a red herring. Whenever a concept changes, it does so by virtue of changes in the beliefs that employ that concept (or predicate, if you are thinking in terms of sentences). For example, if you recategorize whales as mammals rather than fish, you have made an important change in the concept *whale*. But this amounts to no more than deleting the belief that whales are fish and adding the belief that whales are mammals. Your concept of mammal may also change by adding the belief that whales produce milk, but this merely follows from the other belief addition. So as far as epistemology is concerned, conceptual change is redundant with respect to the central question of belief revision.

Thus anyone who thinks conceptual change is important has to give an account of it that goes beyond mere belief revision.

The above argument assumes that the principles according to which beliefs are added and deleted operate independently of considerations of conceptual structure. If you are a Bayesian, belief revision is just a matter of changing probability distributions over the set of propositions (see, for example, Horwich 1982). But suppose that you want to take a more psychologically realistic approach to belief revision, one that could account for why some revisions are harder to make than others and why some revisions have more global effects. Perhaps such facets of belief revision can only be understood by noticing how beliefs are organized via concepts. There may be a difference between deciding that whales are mammals and deciding that whales have fins, a difference that can only be understood in terms of the overall structure of our conceptual system, relating *whale* to *mammal* in ways more fundamental than simply having the belief that whales are mammals. For the moment, this is only a possibility, not a refutation of the argument that conceptual change is just belief revision. But it is enough to suggest that it is worth exploring the cognitive science literature on concepts for suggestions about how conceptual structure could matter to belief revision.

2.3 WHAT ARE CONCEPTS FOR?

Before proceeding further, some clarification is in order concerning concepts and predicates, and sentences and propositions. Sentences are syntactic entities, marks on paper. Among their constituents are predicates such as "whale" in the sentence "Gracy is a whale." In contrast, I shall treat concepts and propositions as mental representations, with concepts corresponding to predicates and propositions corresponding to sentences. In my usage, concepts are mental structures representing what words represent, and propositions are mental structures representing what sentences represent. This mentalistic interpretation is not the only one possible: a Platonist could treat concepts as the meaning of predicates and propositions as the meaning of sentences independent of what is in anybody's head. Instead of discussing abstract meanings, I follow many researchers in psychology and artificial intelligence in supposing that concepts are mental structures analogous to data structures in computers. Anyone who is wedded to concepts and propositions as abstractions should reinterpret my concepts as "conceptual representations" and my propositions as "propositional representations." Whereas words are parts of sentences and not vice-versa, concepts and propositions as mental representations can be parts of each other (see section 2.5).

To prevent additional terminological confusion, it is necessary to point out that researchers in cognitive psychology and artificial intelligence tend to use the terms "knowledge" and "belief" differently from philosophers, who often characterize knowledge as something like true justified belief. Their use of "knowledge" is closer to philosophers' use of "belief." Cognitive scientists have also taken to using the term "epistemology" broadly to cover anything having to do with knowledge in a diluted sense ignoring justification. In this book I generally use "knowledge" and "epistemology" in their traditional philosophical senses that presuppose questions of justification.

Psychologists have many reasons for being interested in the nature of concepts. Whereas the epistemologist's primary concern is with the question of justification, the psychologist must try to account for many different kinds of behavior. Here is a list, undoubtedly incomplete, of various roles that concepts have been deemed to play, using the concept *whale* as an example.

1. *Categorization.* Our concept *whale* enables us to recognize things as whales.
2. *Learning.* Our concept *whale* must be capable of being learned, perhaps from examples, or perhaps by combining other existing concepts.
3. *Memory.* Our concept *whale* should help us remember things about whales, either in general or from particular episodes that concern whales.
4. *Deductive inference.* Our concept *whale* should enable us to make deductive and inductive inferences about whales, for example enabling us to infer that since Gracy is a whale, she has fins.
5. *Explanation.* Our knowledge about whales should enable us to generate explanations, for example saying that Gracy swims *because* she is a whale.
6. *Problem solving.* Our knowledge about whales should enable us to solve problems, for example how to get an errant whale out of the harbor.
7. *Generalization.* Our concept *whale* should enable us to learn new facts about whales from additional examples, for example to form new general conclusions such as that whales have blubber under their skin.
8. *Analogical inference.* Our concept *whale* should help us to reason using similarities: if you know that dolphins are quite intelligent and are aquatic mammals like whales, then whales are perhaps intelligent too. Metaphor should also be supportable by the concept, as when we say that someone had a whale of an idea.
9. *Language comprehension.* Our understanding of sentences such as "Gracy is a whale" depends on our knowing something about the concept *whale.*
10. *Language production.* We need to be able to utter sentences like "Gracy is a whale" and "Whales are less friendly than dolphins."

Ignoring the last two language issues, which introduce problems not directly connected to belief revision, we can examine whether the first eight roles require that belief change pay attention to conceptual structure. Categorization might be seen as a straightforward case of belief application: you believe that any large sea-object that moves and blows water into the air is a whale, so you categorize the large blob in the ocean producing spray as a whale. You thereby add the belief "the blob is a whale" to your set of beliefs. But categorization is rarely so simple as this deduction, since unexceptionable rules are hard to come by. Submarines are also large sea-objects that move and can blow water into the air. So in categorizing the blob as a whale rather than as a submarine you will need to decide which concept fits the blob better, and fitting the concept may be more than a matter of simple belief application (see the discussion of categorization in Holland, Holyoak, Nisbett, and Thagard 1986, ch. 6).

Identifying the blob as a whale presupposes that you have already learned the concept of a whale, but what does this amount to? The belief-revision approach to epistemology never addresses the question of the origin of the concepts (or predicates, if you prefer) that are essential components of beliefs. Without the concept of a whale you could never form the belief that Gracy is a whale. I recently learned the concept of a narwhal, which like a whale is a cetacean (marine mammal with large head and hairless fishlike body), but has a long ivory tusk. Now, with the help of the tusk criterion and a picture of a narwhal I saw in my dictionary, I can potentially form the belief "the blob is a narwhal." But how are such concepts formed? The psychological and computational literature on concept formation suggests two principal ways: by learning from examples and by combining previously existing concepts (see section 3.4). Neither of these is a simple matter of adding new beliefs. In particular, forming new concepts by combining old ones requires that the concept have much structure, because the new concepts are not simple sums of the old (Thagard 1988, ch. 4; Holland et al., ch. 4). To form, for example, the concept *walking whale* we have to decide how to reconcile what we know about whales with what we know about walking, perhaps by concluding that whales merely imitate walking by floating with their tails near the bottom of the ocean and wiggling their fins. Belief acquisition presupposes conceptual organization.

Is memory important for epistemology? For Harman (1986) and Goldman (1986) it professedly is, yet neither considers the role that concepts have been conjectured to play in memory. We need to be able to remember beliefs in order to use them to revise others, and conceptual organization can be highly relevant to memory. If spreading activation of concepts is a crucial way in which beliefs get accessed (Thagard 1988, ch. 2), then the organization of concepts matters to memory and hence to belief revision based on memory. The organization of concepts is also important for theories of access to memory based on semantic similarity of concepts (Thagard, Holyoak, Nelson, and Gochfeld 1990).

The defender of a pure belief-revision approach to epistemology might say that at least the next three roles in my list—deductive inference, explanation, and problem solving—do not require any attention to the conceptual structure. Explanation and problem solving are approximated by deduction, and deduction is just a matter of deriving the consequences of a set of beliefs. In a real system, however, there must be constraints on what gets deduced to avoid the explosion of exponentially increasing numbers of beliefs: a system that expands its data base by inferring from A to A&A, A&A&A, etc., will quickly be swamped. One way of constraining deduction is to draw inferences only from a subset of beliefs deemed to be active, and one way of controlling the activation of beliefs is through the activation of concepts to which they are attached (Thagard 1988). Moreover, some deductive inferences may be performed by processes that directly use conceptual structures rather than typical rules of inference. If the structure for *whale* contains the information that a whale is a kind of mammal, then information stored with the structure for mammal can then be "inherited" by whale. A frame system can infer that whales produce milk by virtue of the *kind* link between the whale frame and the mammal frame. What is inferred is the same as what a logic system would conclude using standard deduction from sentences, but the procedure is more direct. Hence deduction, and perforce explanation and problem solving, may be served by additional conceptual structure.

Generalization, such as inferring from some examples that whales have blubber, may also benefit from conceptual structure. We can infer a generalization from fewer instances if we know something about the variability of the kinds of things under consideration (Holland et al. 1986, ch. 8). For example, a few instances of whales that have blubber under their skin may be enough to convince you that all whales have blubber under their skin, because kinds of mammals, cetaceans, and sea-creatures do not vary much in their subcutaneous attributes. In contrast, if you sec a few whales swimming in circles near volcanoes you will be hesitant to infer that all whales swim in circles near volcanoes. Your background knowledge tells you nothing about the variability of the behavior of mammals, or cetaceans, or sea-creatures near volcanoes. Crucial to such inferences is knowing what kinds of thing whales are, and this could be part of the structure of the concept *whale*.

Finally, analogical inference may well require much conceptual structure for several of the reasons already mentioned. Use of an analog in inference, problem solving, or explanation requires retrieving it from memory, which can depend heavily on the structure of concepts (Thagard et al. 1990). Mapping from one analog to another to determine what corresponds to what can require judgments of semantic similarity that also depend on conceptual structure (Holyoak and Thagard 1989).

None of the remarks just made is conclusive. We do not have a definitive theory covering the areas of cognition mentioned above. But there are enough psychological and computational experiments to suggest that the postulation of conceptual structure may be important for understanding many cognitive phenomena relevant to belief revision. Now let us look at some recent proposals about what concepts might be from Edward Smith, connectionists, and George Miller.

2.4 WHAT ARE CONCEPTS?

Smith (1989) reviews recent experimental and theoretical research in cognitive psychology on concepts, starting with the traditional view that concepts can be defined by giving necessary and sufficient conditions for their application. This view has two major problems: (1) it is nearly impossible to find defining conditions for nonmathematical concepts, and (2) concepts show typicality effects. Apples and peaches, for example, are more typical fruits than figs or pumpkins. Typicality effects have led psychologists to consider concepts in terms of *prototypes*, or best examples. Something is categorized as a fruit if it is sufficiently similar to our prototype of a fruit, and typicality is a function of degree of similarity to the prototype (see also Lakoff 1987 for a discussion of prototypes).

The account of concepts as prototypes has, however, encountered some theoretical and empirical difficulties of its own. Armstrong, Gleitman, and Gleitman (1983) have found typicality effects even in crisp mathematical concepts: 3 is a more typical odd number than 359. Smith argues that a concept includes a *core* as well as a prototype, where the core is more diagnostic than the prototype even though the prototype may be useful for quickly identifying instances. Thus the core of *odd number* would be defined mathematically in terms of divisibility, even if we quickly decide that something is an odd number by matching it against prototypes. In nonmathematical concepts, cores will not be strict definitions, but will nevertheless serve to give concepts stability. Following Murphy and Medin (1985), Smith observes that there is more to categorization than simply matching to a prototype, since causal reasoning can play a role. For example, if you see a man at a party jump into a swimming pool, you may categorize him as drunk, not because jumping into swimming pools matches your prototype drunk, but because the hypothesis that he is drunk explains his behavior. As I suggested above, explanation is one of the roles that concepts must serve, so more than the prototype or prototype + core theories is needed. (See also Medin 1989; Ahn and Brewer 1991; Kunda, Miller, and Claire 1990.)

Computational implementations of prototypes have centered on Minsky's (1975) notion of a frame. Frames are symbolic structures that specify for a concept various slots and default values for a slot. For example, a frame representing the concept *whale* would have a slot *size* with the value *large*. Default values are not definitional but merely express typical expectations. Recently, a different kind of computational approach has been proposed by connectionists, who theorize using networks modeled loosely on the brain. As we saw in section 2.1, Rumelhart et al. (1986) advocate a view of schemas (concepts) as patterns of activation distributed over neuron-like units in a highly connected network. A concept is not a structure stored in the brain the way a data structure is stored in a digital computer. Rather, it emerges when needed from the interaction of large numbers of connected nodes.

The currently most influential learning technique for producing distributed representations is *back propagation*, in which a trainer tells the output units of a network whether they got the right answer and weight adjustments are propagated back through the network to make it more likely that the right answer will be generated in subsequent runs. Figure 2.2 provides an abstract

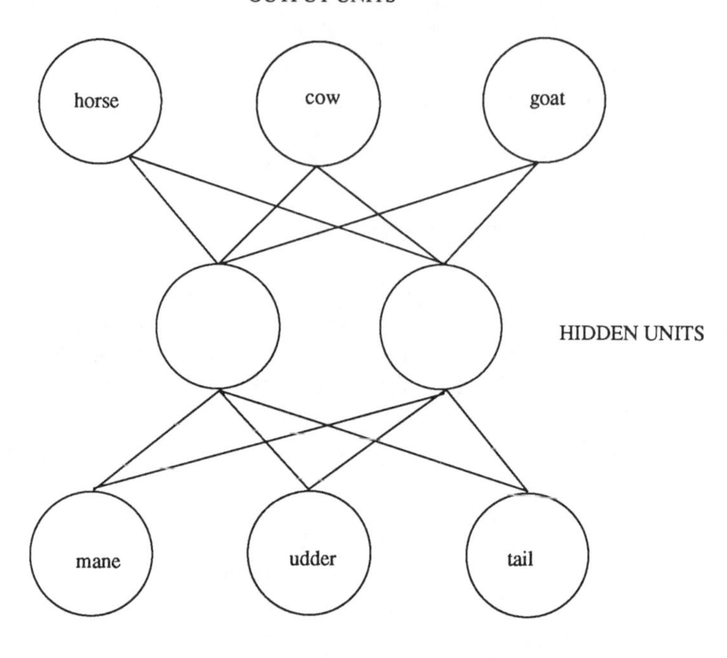

Figure 2.2. A simple network with three layers. Circles represent units, while the straight lines are links between the units. Each link has a weight representing the strength of the connection.

example of a simple three-layer network. The circles represent units that can have activation values representing the presence or absence of features. Suppose, for example, that the purpose of the network is to discriminate between horses, goats, and cows. Features of sample animals are presented to the network by activating input units, for example ones for having a mane and having a tail. The top output layer which indicates the network's decision could be interpreted as identifying a given animal as one of these, depending on which units are on or off. The network is *trained* by presenting samples of features to the input units and having a critic detect whether the correct output units are being turned on. For example, if the inputs activate the units for mane and tail, then the unit horse should be active. If, however, a mistake is made, then the back propagation algorithm is used to adjust the weights on the links (shown by straight lines) between the output layer and the hidden units, and between the input layer and the hidden units. With sufficient iterations, the network will be able to identify which animal goes with which features. This information is not stored in particular rules, but rather is distributed over the connections and activation patterns that occur in the network. Distributed representations are contrasted with *local* representations in which a single node in a network represents an entire concept or proposition. An anticipation of the view that knowledge is not embedded in rules can be found in Kuhn (1977, 312).

The connectionist view of concepts appears promising for accounting for subtle categorization effects. A network could, for example, acquire the concept of a whale by being trained on examples of whales, learning to identify blobs as whales without acquiring explicit slots or rules that state typical properties of whales. No single unit corresponds to the concept *whale*, since information about whales is distributed over numerous units. Work in progress is investigating how concepts as learned patterns of activation can even be organized into kind-hierarchies and part-hierarchies (Hinton 1988). Memory also appears tractable from a connectionist viewpoint, since retrieving information about a concept should happen automatically when the right pattern of activation arises. Another appealing aspect of the connectionist approach is that it in principle has no problem with partially nonverbal concepts such as *red* or *the taste of gorgonzola cheese*, since these also can be patterns of activation acquired by training.

Nevertheless, connectionists cannot be said to have a full theory of concepts, since no account has been given of:

1. how concepts can be formed by combination rather than from examples;
2. the use of concepts in relatively sequential processes such as explanation, problem solving, and deduction;
3. how distributed representations of concepts can be used in a wide range of inferential tasks including complex generalizations and analogies; and
4. how distributed representations can be used in a general theory of language production and comprehension.

Perhaps progress will come on all these fronts, but for now we cannot rely only on connectionist ideas for help in understanding the nature of concepts. Although connectionists show us how to understand prototypes as patterns of activation derived from examples, they have a daunting task in showing how such concepts can generate explanations as in the above case of the drunk in the swimming pool.

A different although perhaps ultimately complementary view comes from the work by George Miller and his colleagues on the structure of the mental lexicon (Miller et al. 1990; Miller and Johnson-Laird 1976). WordNet is a large electronic lexical reference system based on psycholinguistic theories of the organization of human lexical memory. A concept is represented by a set of synonyms, and synonym sets are organized by kind, part, and antonymy relations. Kind-relations and part-relations are fundamental to the organization of the lexicon because they generate hierarchies. For example, a whale is a kind of cetacean, which is a kind of mammal, which is a kind of animal, which is a kind of living thing. A toe is part of a foot, which is part of a leg, which is part of a body. Kind-hierarchies and part-hierarchies serve to structure most of our conceptual system, providing backbones off which other conceptual relations can hang. Although well known to psycholinguists, the importance of these hierarchies has been neglected by philosophers who have tended to speak of "conceptual schemes" entirely in the abstract. (See also Cruse 1986 for more on hierarchies, and see Winston, Chaffin, and Herrman 1987 for more on part-relations.)

WordNet now includes many thousands of entries, including verbs and adjectives as well as nouns. One advantage of working on such a large scale is that the differences between the kinds of lexical items become readily apparent. Kind-hierarchies and part-hierarchies apply well to nouns, but adjectives are primarily organized into antonymic clusters such as that posed by the extremes *wet-dry*. According to Miller et al., verbs do not form kind-hierarchies or part-hierarchies, but are organized in terms of relations of manner and entailment (but see section 9.3.2). For example, to nibble is to eat in a specific manner, and driving entails riding. Whereas most research on the nature of concepts has been restricted to nouns, a general theory should attend to other parts of speech as well. Later chapters will reveal, however, that revolutionary conceptual change largely involves concepts represented by nouns.

WordNet's semantic relations are not intended to exhaust the meaning of a concept. Miller and Johnson-Laird (1976) have a procedural theory of meaning that associates concepts with computational routines, often tied to perception, for identifying instances of a concept. But they say (p. 696) that semantics also requires placing concepts and sentences in the context of a larger system of knowledge and belief. I presume this larger context would include information of the sort that can be used for generating explanations and performing the other roles of concepts discussed above. Conceptual information should therefore be tied to world knowledge such as that drunks tend to act wildly, which, with the characterization of jumping into the swimming pool as acting wildly, could prompt the explanatory hypothesis that the man in the swimming pool was drunk. I shall now propose how this kind of knowledge can be integrated with other considerations of conceptual structure.

2.5 CONCEPTS AS COMPLEX STRUCTURES

No one currently knows how concepts are stored in the brain. Perhaps they are patterns of activation of neurons as the connectionists suggest, or maybe some more complex organization and distribution exists. Without worrying about neural implementation, we can nevertheless consider how concepts are organized and serve to play the numerous roles listed above. I am not saying that connectionism and neuroscience are irrelevant: we are still in the early stages of cognitive science and should not tolerate imperialistic limitations on kinds of approaches. My proposals will be at the level of traditional symbolic artificial intelligence, but are meant to suggest targets for subsequent connectionist and neurological analysis. In contrast to the sometimes acrimonious debate between proponents of symbolic AI and purportedly subsymbolic connectionism, I see cognitive science as using a continuum of complementary computational methods (see section 9.3.4 for further discussion).

What is the concept of a whale? Let us start with WordNet-style lexical organization. Information about *whale* could include something like the following:

WHALE:
A kind of: cetacean, mammal, sea-creature.
Subkinds: humpback, blue, killer, sperm, white, beluga, etc.
Parts: fins, blubber, bone, blowhole, tail.

A WordNet entry can in addition include lists of synonyms, antonyms, and wholes. For a full representation of the concept, we can supplement this representation with pointers to individual whales like Gracy that constitute the known instances of whales. The concept of whale should, in addition, provide access to various general facts about whales that are important for conversing and reasoning about whales. It should therefore be connected with rules such as the following:

R1. If x is a whale, then it swims.
R2. If x is a whale, and x surfaces, then x blows air through its blowhole.

Holland et al. (1986) discuss the relevance of rules for concepts. Notice that R1 is easily represented as a slot in a Minsky-style frame:

WHALE:
locomotion: swims.

In contrast, R2 with more complicated conditions and the complex relation *blows* is less amenable to treatment as a slot. It would also pose problems for simple connectionist learning schemes that form patterns of activation merely from features. The word "whale" is unusual in lacking synonyms and antonyms, which abound especially for adjectives.

My proposal then is to think of concepts as complex structures akin to frames, but (1) giving special priority to kind and part-whole relations that establish hierarchies and (2) expressing factual information in rules that can be more complex than simple slots. Schematically, a concept can be thought of as a frame-like structure of the following sort:

CONCEPT:
A kind of:
Subkinds:
A part of:
Parts:
Synonyms:
Antonyms:
Rules:
Instances:

If concepts are computational structures of the sort just described, then it makes sense to say that rules are parts of concepts as well as that concepts are parts of rules. The presence of rules shows how concepts can be used in deduction, explanation, and problem solving. It also shows how concepts can be intimately tied in with theories, since some rules express causal relations. A slightly sophisticated understanding of whales involves an explanation of blowholes in terms of whales' getting oxygen from the air, not from water like fish, so that the function of the blowhole is to expel air. Concepts as complex structures can figure in generalization, analogy, and other forms of inference (Thagard 1988).

We are still far from having a full theory of concepts, but the enriched account offered so far makes possible a start on considering how conceptual structure can be relevant to epistemic change. I shall show in the next chapter how to distinguish different kinds of epistemic change that would be opaque from the point of view of belief revision alone.

2.6 CONCEPTUAL HIERARCHIES

An understanding of conceptual revolutions requires much more than a view of the nature of isolated concepts. We need to see how concepts can fit together into *conceptual systems* and what is involved in the replacement of such systems. On the view of concepts just presented, conceptual systems consist of concepts organized into kind-hierarchies and part-hierarchies and linked to each other by rules.

In this spirit, a conceptual system can be analyzed as a network of nodes, with each node corresponding to a concept, and each line in the network corresponding to a link between concepts. For example, Figure 2.3 provides a small fragment of a conceptual network about animals. The concept nodes are marked by names in ellipses. This network uses five kinds of links:

1. *Kind* links, marked by straight lines labeled "K." These links indicate that one concept is a kind of another: for example, canary is a kind of bird and bird is a kind of animal.

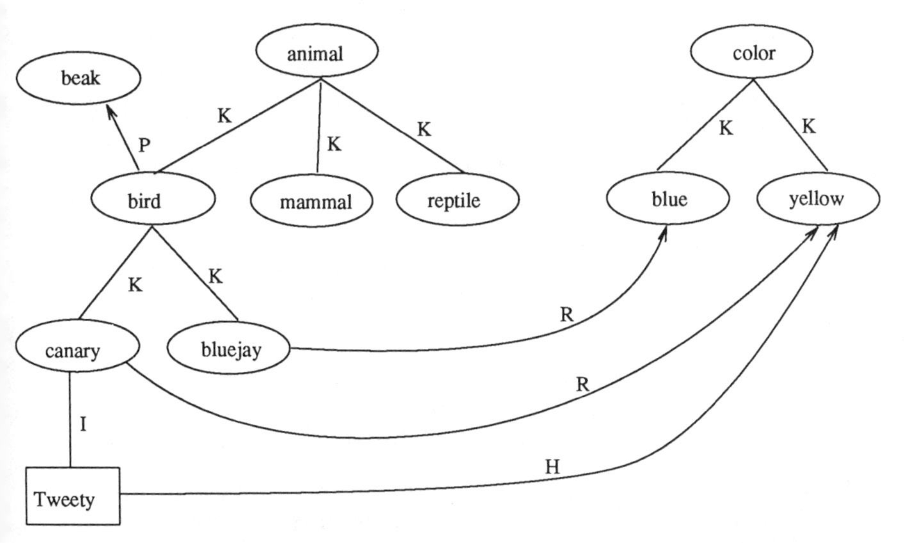

Figure 2.3. Example of a conceptual hierarchy. Kind links are marked by straight lines labeled "K." Instance links are marked by straight lines labeled "I." Rule links are marked by curved lines terminating with arrows and labeled "R." Property links are marked by curved lines terminating with arrows but labeled "H" for "has-property." Part-whole links are marked by straight lines terminating with arrows and marked "P."

2. *Instance* links, marked by straight lines labeled "I." These indicate that some particular object, marked by a box rather than by an ellipse, is an instance of a concept: Tweety is a canary. The chain of links in the network shows that Tweety is also an animal.
3. *Rule* links, marked by curved lines terminating with arrows and labeled "R." These express general (but not always universal) relations among concepts, for example that canaries have the color yellow.
4. *Property* links, marked by curved lines terminating with arrows but labeled "H" for "has property." These indicate that an object has a property such as: Tweety is yellow.
5. *Part* links, marked by straight lines terminating with arrows and marked "P." These indicate that a whole has a given part: a beak is a part of a bird.

Relations and higher-order predicates can also be expressed within conceptual networks, since we can draw lines to indicate, for example, that yellow is brighter than blue.

The logician will immediately notice that Figure 2.3 contains no information that cannot be represented in first-order predicate calculus. Kind links and rule links can be encoded by universal generalizations such as **(x)(canary(x) → bird (x))**, and instance links and property links can be encoded by atomic sentences such as **bird(Tweety)**. But from a computational perspective the links have importance that transcends the information that they express: they can make possible *procedures* different from those associated with predicate calculus. Thus although AI knowledge representation systems may be *expressively* equivalent to logic-based systems, they need not be *procedurally* equivalent. The distinction between expressive and procedural equivalence of systems and the claim that AI techniques go beyond logic systems are defended elsewhere (Thagard 1988, 1984).

If a conceptual system consists of a network of nodes with links such as those just described, then conceptual change consists of adding or deleting nodes and links. The most common changes involve the addition of:

1. New concept nodes. Example: new concept *ostrich.*
2. New kind links between concept nodes. Example: canaries are reptiles.
3. New rule links between concept nodes. Example: canaries have color blue.

Changes often come in groups rather than in isolation. For example, addition of the concept node for ostrich will be accompanied by addition of a kind link from *ostrich* to *bird*, addition of instance links from *ostrich* to various object nodes representing particular ostriches, and addition of rule links between *ostrich* and concept nodes such as *long-legged.*

But not all additions and deletions are equally serious. The most dramatic changes involve the addition of new concepts as well as new rule and kind links, where the new concept and links *replace* ones from the old network. We can distinguish replacement of a conceptual system from simple deletions and additions if some of the previous links remain, indicating that the new concepts and links have a place in the new system similar to those in the old system. Although a conceptual revolution can involve dramatic replacement of a substantial portion of a conceptual system, continuity is maintained by the survival of links to other concepts. Dramatic changes will be most visible in the hierarchies of concepts that can be built out of kind links and part links. These hierarchies provide a scaffold that arranges and organizes other concepts. Hence changes in kind-relations and part-relations usually involve a restructuring of conceptual systems that is qualitatively different from mere addition or deletion of nodes and links. The next chapter contains much more about how conceptual hierarchies affect conceptual change.

Why are kind-relations and part-relations so fundamental to our conceptual systems? In addition to the organizing power of the hierarchies they form, these two sets of relations are important because they specify the constituents of the world. *Ontology* is the branch of philosophy (and cognitive science!) that asks what fundamentally exists, and ontological questions usually concern what *kinds* of things exist. Moreover, given an account of the kinds of things there are, which translates immediately into a hierarchical organization, we naturally want to ask: of what are the objects of these kinds made? The answer to this question requires consideration of their parts, generating the part-hierarchy that also organizes our concepts. Thus the major role that kind-hierarchies and part-hierarchies play in our conceptual systems is not accidental, but reflects fundamental ontological questions.

Knowledge of parts is relevant to knowledge of kinds. B. Tversky (1989) argues that children's knowledge of parts makes an important contribution to their ability to divide things into kinds. Things that have the same parts can be judged similar in both appearance and function and therefore can naturally be classified as being of the same kind. Similarly, Markman (1989) contends that when children learn the meaning of new words, they implicitly assume that a word applies to whole objects of a particular kind. Lewis (1990) shows how set theory can be reformulated in terms of part-relations: dogs are a subset of animals if and only if the set of dogs is part of the set of animals. We cannot, however, identify kinds with sets, since any collection of objects can form a set, whereas people are discriminating about how they group things into kinds.

Although I hope to have established the importance of conceptual hierarchies, the case against the view that all epistemic change is belief revision is still incomplete. It requires much more detailed discussion of conceptual change found in the next chapter.

2.7 SUMMARY

Theorists in philosophy, psychology, and artificial intelligence have proposed different views of the nature of concepts. A rich account of concepts and conceptual change is needed to overcome the widely held view that the growth of scientific knowledge can be understood purely in terms of belief revision with no reference to conceptual change. Concepts serve many psychological functions, and can be understood as complex computational structures organized into kind-hierarchies and part-hierarchies. Such structures also involve rules that can contribute to explanations.

CHAPTER 3

Conceptual Change

USING the view of conceptual organization developed in the last chapter, we can rank the severity of different kinds of epistemic change. The first part of this chapter provides such a ranking, in which changes involving kind-relations and part-relations are especially important. Much more important than a ranking of conceptual changes is showing that the most dramatic sorts have occurred in major scientific episodes. Hence the middle part of this chapter gives an analysis of the developments that occurred in the chemical revolution, when the phlogiston theory of combustion was replaced by Lavoisier's oxygen theory. We shall see that this revolution involved substantial systematic changes in both kind-relations and part-relations, as well as introduction of new concepts and deletion of old ones. The latter part of this chapter outlines a theory of conceptual development adequate to account for the case of Lavoisier and for the other scientific revolutions discussed in later chapters.

3.1 DEGREES OF CONCEPTUAL CHANGE

It would be futile to try to offer criteria for identity of concepts that attempt to specify when a concept ceases to be the concept it was. We cannot even give such criteria for mundane objects like bicycles: if I change the tires on my bicycle, is it the "same" bike? What if I change the wheels, or the frame, or all of the above? But without giving a definition of sameness for bicycles, we can nevertheless rank degrees of change. Replacement of the parts mentioned are all changes in my bicycle, but it seems clear that changing the frame is a more severe change than the wheels, which is more severe than changing the tires. Similarly, we can characterize different kinds of conceptual change and see that some are more serious than others.

Table 3.1 provides a list of kinds of conceptual change, roughly ordered in terms of degree of increasing severity. Although the first few items on the list can be fully understood as belief revision, later items involve conceptual changes whose epistemic import goes beyond belief revision. The list considers additions to conceptual structure, but could easily be expanded to include deletions too. The first kind of change, adding a new instance, involves a change to the structure of the concept *whale*, but is relatively trivial, like adding a pennant to a bicycle. Equally trivial is the deletion consisting of

Table 3.1
Degrees of Conceptual Change

1. Adding a new instance, for example that the blob in the distance is a whale.
2. Adding a new weak rule, for example that whales can be found in the Arctic ocean.
3. Adding a new strong rule that plays a frequent role in problem solving and explanation, for example that whales eat sardines.
4. Adding a new part-relation, for example that whales have spleens.
5. Adding a new kind-relation, for example that a dolphin is a kind of whale.
6. Adding a new concept, for example *narwhal.*
7. Collapsing part of a kind-hierarchy, abandoning a previous distinction.
8. Reorganizing hierarchies by *branch jumping*, that is, shifting a concept from one branch of a hierarchical tree to another.
9. *Tree switching*, that is, changing the organizing principle of a hierarchical tree.

deciding that the blob is not a whale after all. The second and third kinds of change involve adding rules of different strength. The terms "weak" and "strong" indicate the importance of the rule for problem solving. Thus the distinction between change 2 and change 3 is pragmatic: if you are an Eskimo or Russian fisherman, 2 might be a stronger rule.

Adding new part-relations, change 4, can be called *decomposition.* Conceptual change typically occurs in the part-hierarchy when new parts are discovered, as has often occurred in physics with the discovery of molecules, atoms, and subatomic particles. Particularly dramatic are decompositions of what were previously thought to be indivisible wholes, such as water before Lavoisier and the atom before Thomson discovered it contains electrons. Change 5 involves adding kind-relations. Carey (1985) discusses *coalescence*, adding a new superordinate kind that links two things previously taken to be distinct, and *differentiation*, making a distinction that produces two kinds of things. She illustrates coalescence by the conceptual change that occurs in children when they form the concept of *alive* that has as subordinates both animals and plants (see section 10.2.2 for further discussion). Her prime example of differentiation concerns heat and temperature, which were taken to be the same until Black distinguished them.

The additions outlined in changes 1–5 only become possible when one has formed a concept of whale as a distinct kind of entity. Concepts can be added for a variety of reasons, including coalescence. In the nineteenth century, scientists realized that electricity and magnetism were fundamentally the same and produced the coalesced concept of electromagnetism. New concepts can also be introduced for explanatory reasons, for example when the concept *sound wave* was formed as part of the explanation of why sounds behave as they do (Thagard 1988). Concepts such as *oxygen*, *electron*, *quark*, and *gene* were all introduced as parts of explanatory theories. Thus change 6, adding concepts, is an important part of the development of scientific knowledge.

Science often develops by introducing new distinctions in the form of kind-relations, as in change 5. But conceptual change can also involve the attenuation or abandonment of distinctions previously made, producing a *collapse* of part of the kind-hierarchy. This is change 7 in Table 3.1; collapse is the reverse of differentiation. Darwin attenuated the distinction between species and varieties (see Chapter 6), and Newton completely abandoned the Aristotelian distinction between natural and unnatural motion (see Chapter 8).

Because kind-relations organize concepts in tree-like hierarchies, a very important kind of conceptual change, 8 in Table 3.1, involves moving a concept from one branch of the tree to another. Such *branch jumping* is common in scientific revolutions. For example, the adoption of Copernican theory required the reclassification of the earth as a kind of planet, when previously it had been taken to be *sui generis*. Similarly, Darwin recategorized humans as a kind of animal, when previously they were taken to be a different kind of creature. I shall provide examples of branch jumping in all the major scientific revolutions.

Change 9, affecting the organizing principle of a hierarchical tree, is the most dramatic kind of conceptual change. Darwin not only reclassified humans as animals, he changed the meaning of the classification. Whereas before Darwin *kind* was a notion primarily of similarity, his theory made it a historical notion: being of common descent becomes at least as important to being in the same kind as surface similarity. Einstein's theory of relativity changed the nature of part-relations, by substituting ideas of space-time for everyday notions of space and time. Later chapters will illustrate these changes in organizing principles at much greater length.

If one does not attend to conceptual structure, and thinks only in terms of belief revision, the importance of the last five kinds of conceptual change will be missed. Although 1–3 can be interpreted as simple kinds of belief revision, 4–9 cannot, since they involve conceptual hierarchies. In particular, branch jumping and tree switching are changes that are very difficult to make on a piecemeal basis. Darwin did not simply pick away at the creationist conceptual structure: he produced an elaborate alternative edifice that supplanted it as a whole. Adopting a new conceptual system is more holistic than piecemeal belief revision, as will be shown more thoroughly later in this chapter.

To give a slightly different perspective, Figure 3.1 provides a taxonomy of different kinds of epistemic change. Straightforward belief revision involves the addition or deletion of beliefs. Conceptual change goes beyond belief revision when it involves the addition, deletion, or reorganization of concepts, or redefinition of the nature of the hierarchy. Simple conceptual reorgani-

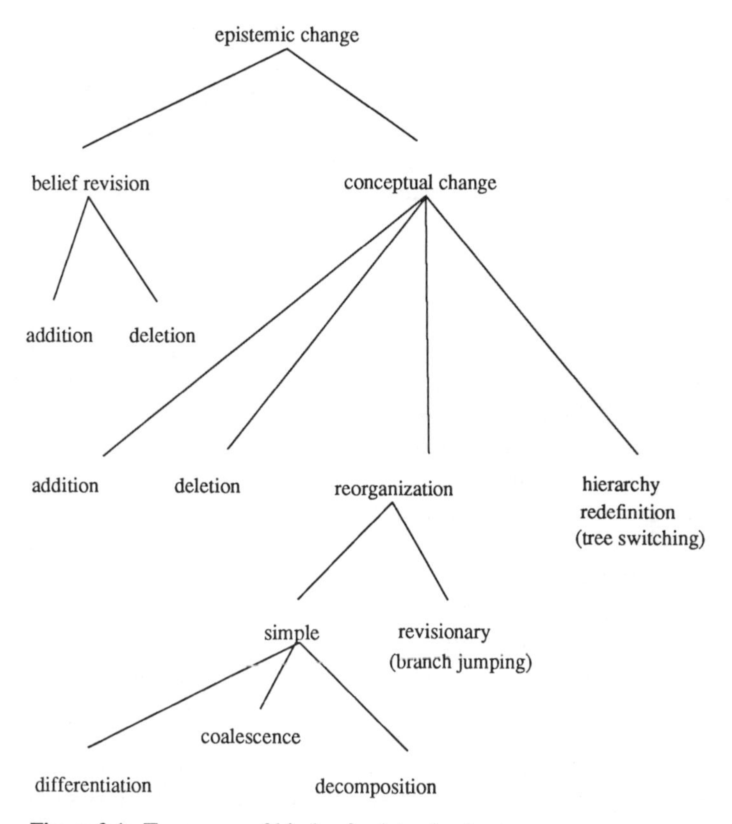

Figure 3.1. Taxonomy of kinds of epistemic change.

zation that involves mere extension of existing relations differs from the revisionary sort represented by branch jumping, which involves moving concepts around in the hierarchies and rejecting old kind-relations or part-relations as well as adding new ones. Belief revision, concept addition, and simple reorganization of conceptual hierarchies are common in the development of scientific knowledge, but we shall see that branch jumping and tree switching are much rarer events associated with conceptual revolutions. The momentousness of a revision is affected, of course, by more than these conceptual relations. Some beliefs are important to us because they are closely related to our personal goals, while other beliefs are important because they are densely related to many other beliefs (see section 4.4).

I have been primarily illustrating conceptual change so far using the concept *whale*, but now want to consider a scientific example that is embedded in a richer theoretical context. The concept of an *acid* has undergone dramatic changes in its long history (Walden 1929). The originating Latin term (*acidus*) just means sour, and was applied to things like vinegar and lemon juice; the Latin word for vinegar, *acetum*, comes from the same root as *acidus*. What we now think of as the acid components of vinegar and lemon juice, acetic acid and citric acid, were not identified until much later. Only in the Middle Ages did alchemists isolate what we now call nitric acid and sulfuric acid. The French term *acide* arose in the sixteenth century and the English term "acid" in the seventeenth, as an adjective meaning sour but also as a noun vaguely denoting a class of sour substances. It seems that the earliest concept of an acid was for sour substances whose subkinds included vinegar and nitric and sulfuric acids.

As far back as Democritus, however, the attempt was made to explain *why* some things are sour. He offered the atomistic explanation that how something tastes is caused by the shape of the atoms that make it up (Sambursky 1956, 122). Similarly, in the seventeenth-century Nicholas Lémery proposed that the properties of acids derive from the sharp, spikey form of the corpuscles that make them up. In the phlogiston theory, which dominated mid-eighteenth-century chemistry, acids were taken to be simple substances while what we now conceive of as their constituents were understood as compounds. For example, sulfur was thought to be a compound of oil of vitriol (sulfuric acid) and phlogiston.

The first general theory of acids originated with Lavoisier, who thought that the central property of acids was that they contain oxygen. In fact, "oxygen" derives from Greek words meaning "sour producing." As we will see in the next section, Lavoisier abandoned the concept of phlogiston, and viewed acids as compounds rather than simple substances: sulfuric acid is a compound of sulfur, oxygen, and hydrogen. For Lavoisier, oxygen was more than just a constituent of acids: it was the *principle* of acidification that gave acids their sourness and other central properties.

Lavoisier's theory of acids was demolished, however, in 1815 when Humphrey Davy showed that muriatic acid—what we now call hydrochloric acid—consists only of hydrogen and chlorine. He and Liebig contended that hydrogen, rather than oxygen, was essential to the constitution and effects of acids. This hypothesis was made quantitative by Arrhenius in 1887 when he proposed that acids are substances that dissociate to produce hydrogen ions. More general conceptions of acids were proposed by Brønsted, Lowry, and Lewis in the 1920s. The Brønsted-Lowry account characterizes an acid as a substance that donates protons, so that having hydrogen as a part is no longer essential. Lewis's conception counts still more substances as acids, since acids are characterized as any substances that can accept an electron pair. Chemistry textbooks typically present all three of the Arrhenius, Brønsted-Lowry, and Lewis accounts as useful approximations. Each theory has correlative accounts of the nature of bases and salts. There is no attempt to state a rigid definition of what acids are.

For the understanding of conceptual change, several stages in this development are interesting. Sometime around 1700 the modern concept of acid as a special class of substances with properties other than sourness came into use. *Acid* and *sour* became differentiated. Important changes have taken place in the part-relations of the concept, from the idea that acids have sharp atoms, to the idea that oxygen is their most important part, to the idea that hydrogen is essential, to current ideas that describe acids in terms of protons and electrons rather than elements. The rule that acids typically have hydrogen as a constituent is far more important than the observation, dating from at least the seventeenth century, that acids turn litmus red. The litmus rule has been useful to generations of chemistry students, but plays no role in locating the concept of acid in conceptual hierarchies or in generating explanations of the experiments involving acids.

Over the centuries, there has been a dramatic increase in the number of substances counted as acids, from the few known to the medievals to well over a hundred counted by modern chemists. One interesting consequence of the modern theories is that water can be counted as both an acid and a base, since it can either donate or accept protons and electrons depending on the substance with which it combines. Also of interest is the subclassification of acids, adding important new subkinds such as amino acids. What has occurred in the development of the concept of acid is clearly far more complicated than the refinement of necessary and sufficient conditions for the term "acid": definition and theory go hand in hand. Merely talking of belief revision would obscure the fact that the concept of acid has changed remarkably in its structural relations to other concepts while enjoying a certain stability: we still count vinegar and lemon juice as acidic.

My approach to the problem of conceptual change is based on the organization of mental representations. In contrast, Kitcher's (1988) approach is concerned with the reference of terms. On his view, failures of communication in scientific disputes arise from different ways of fixing the referents of key theoretical terms. Our two approaches are probably compatible, but my concern in this book is with the organization of concepts, not with how or whether they refer to things in the world.

Only a meager understanding of conceptual change can be gained from considering the changes undergone by an isolated concept. Let us now examine an important example of how an entire system of concepts can be replaced.

3.2 THE CHEMICAL REVOLUTION

The supersession of the phlogiston theory by the oxygen theory is a superb example of radical conceptual change. In 1772, when Lavoisier first began to form his views, the dominant theory in chemistry was the phlogiston theory of Stahl (1730). By 1789, when Lavoisier published his *Traité élémentaire de chimie* (Lavoisier 1789), the vast majority of chemists had gone over to Lavoisier's oxygen theory, which gave a very different explanatory account of the phenomena of combustion, calcination, and respiration than the phlogiston theory had provided. Whereas Lavoisier's theory held that combustion, calcination (e.g. rusting), and respiration all involved the absorption of oxygen, the phlogiston theory maintained that these processes all involved the removal of phlogiston. I shall describe the conceptual structure of Stahl's system and of Lavoisier's at four different stages in its development. This depiction reveals the kinds of changes needed to construct a new conceptual system. I shall offer the beginning of an account both of how Lavoisier developed his revolutionary system and of how it supplanted the phlogiston system.

3.2.1 Stahl's Phlogiston Theory

A full account of the chemical revolution would chart every stage in the development of ideas from phlogiston theories to Lavoisier's most developed oxygen theory. I shall concentrate, however, on what seem to be the most important stages in the development of the ideas, providing a kind of snapshot of the conceptual structures at particular times. These snapshots will necessarily depict only part of the organization of the conceptual systems, focusing on the fragments most important to the phlogiston and oxygen theories, namely those concerned with combustion and calcination (oxidation).

The phlogiston theory originated with Georg Stahl, although he drew heavily on ideas of his teacher Johann Becher (Partington 1961). Both rejected the traditional Aristotelian view according to which there are just four elements: earth, air, fire, and water (see section 8.1.1). Figure 3.2 presents a fragment of Stahl's conceptual system, primarily based on Stahl (1730) but depending also on Partington (1961) and Leicester and Klickstein (1952). Stahl says that bodies can be divided into those that are simple principles or those that are compounded. The English word "principle" here is misleading, since, unlike the Latin *principium* and the French *principe* it does not immediately connote something that is basic and indivisible. Stahl's principles are roughly like our elements, basic substances out of which compounds are made, although some are defined more in terms of active function than substance. The simple principles include water and earths, of which there are three kinds: the vitrifiable principle, the liquifiable principle, and the inflammable principle, or phlogiston. A "mixt" is a body that consists of simple principles, whereas a compound may consist of mixts. The properties of compounds are explained in terms of the principles that they contain. Sulfur, for example, burns because it contains the inflammable principle, phlogiston. In Figure 3.2, the rule that

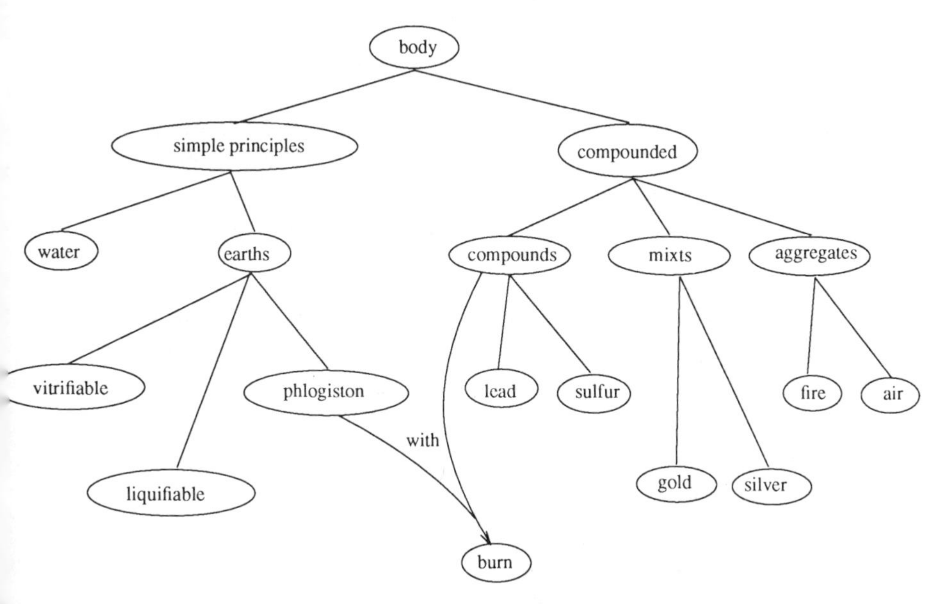

igure 3.2. Fragment of Stahl's conceptual system in 1723. Straight lines indicate kind-lations. The branched curved line with arrowhead indicates a rule.

compounds with phlogiston in them burn illustrates how propositions can link concepts from different parts of the kind-hierarchy. For the sake of legibility I have omitted numerous other links, for example the part-whole links that would indicate that sulfur consists of vitrifiable, liquifiable, and phlogiston principles.

A fuller depiction of Stahl's conceptual system would also show the rule links that enabled the phlogiston theory to explain calcination and respiration. In contrast to our current view that metals combine with oxygen to produce metallic ores (oxides), the phlogiston theory said that ores are simpler than metals. When ores are heated, the phlogiston from the burning charcoal combines with them to produce metals. Calcination, such as the rusting of iron, is the result of the metals losing their phlogiston. Respiration has the effect of removing phlogiston from the body into the air, so that if air is saturated with phlogiston by combustion or breathing, further respiration becomes impossible (Conant 1964). Stahl's conceptual system was thus very broad and provided an explanatory framework for many important phenomena. Despite its radical differences from our current chemical system, which is largely based on Lavoisier's work, we should appreciate the power and comprehensiveness of Stahl's system as well as its strangeness. Stahl discussed, for example, the philosopher's stone of the alchemists. His conceptual system suggests that the convertibility of compounds like lead into mixts like gold should be merely a matter of getting the right combination of principles.

3.2.2 Lavoisier, 1772

According to Guerlac (1961), Lavoisier's interest in combustion began in 1772, when he was twenty-nine years old. The phlogiston theory dominated chemical discussion both on the continent and in Britain. I shall describe how Lavoisier moved farther and farther away from the phlogiston theory, arriving in the 1780s at a highly developed theory of oxygen that was the basis for modern chemistry. My goal here is not to provide a historical narrative, but rather to analyze the conceptual structure of Lavoisier's developing views (for historical accounts, see Holmes 1985, and Donovan 1988). The analysis is a preliminary to developing a computational account of how the full-fledged conceptual system Lavoisier had in 1789 could have evolved. The historical record suggests looking at the following stages:

1. Lavoisier's very early views, circa 1772.
2. Lavoisier's developing views, circa 1774.
3. Lavoisier's developed views, circa 1777.
4. Lavoisier's mature oxygen theory, circa 1789.

Concentration on these stages makes it possible to observe Lavoisier advancing from a vague idea that air might be absorbed in calcination and combustion to a robust alternative to Stahl's theory.

Guerlac (1961) argues that two key phenomena were the source in 1772 of Lavoisier's interest in combustion and calcination and of the view that oxygen combines with substances in both these processes. The first was that when metals are placed in acids, effervescence occurs, which suggested to Lavoisier that air contained in the metals was being given off. In a note written in August of 1772, Lavoisier writes: "An effervescence is nothing but a disengagement of the air that was somehow dissolved in each of the bodies" (Guerlac 1961, 214, my translation). The second source, for which Guerlac has more circumstantial evidence, was the work by Guyton de Morveau that showed conclusively for the first time that objects gain weight in calcination. In another note of August 1772, Lavoisier asserts that "a crowd of experiments appear to show that air enters much into the composition of minerals" (Guerlac 1961, 215, my translation).

Figure 3.3 portrays what I conjecture to be the relevant portion of Lavoisier's conceptual system at that time. Air, calxes, and metals are three kinds of substances. Curved lines represent the rules that effervescing calxes (later: oxides) produce air and that metals gain weight when they become calxes. To explain these rules, Lavoisier conjectures that calxes might contain air, forming the rule labeled "contain?". The dotted lines show an explanatory relation, indicating that the hypothesis that calxes contain air might serve to explain both why calxes effervesce and why metals gain weight in calcination. Figure 3.3 thus displays three kinds of organization: concepts are connected by kind-

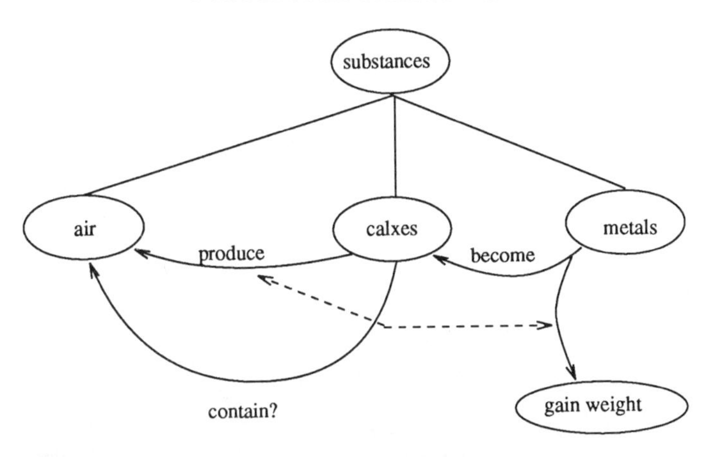

Figure 3.3. Fragment of Lavoisier's conceptual system in 1772. Straight lines indicate kind-relations. Curved lines with arrowheads indicate rules. The dotted lines indicate explanatory relations between rules.

relations and by rules, and rules are connected by explanatory relations (see Chapter 4). Lavoisier did not possess a theory of oxygen at this point, only a vague idea that the presence of air in minerals might explain some puzzling phenomena.

3.2.3 Lavoisier, 1774

Lavoisier quickly moved in September of 1772 to his celebrated experiments on the combustion of phosphorus and sulfur, discovering that in combustion they gain weight. In a sealed note delivered to the French Academy of Sciences in November 1772, Lavoisier said that "this discovery seems to me one of the most interesting since Stahl" (Guerlac 1961, 228, my translation), and suggests that the increase of weight in calcination and combustion may have the same cause: the addition of air.

In January of 1774 Lavoisier published a detailed report of relevant experiments by him and others in his *Opuscules Physiques et Chymiques* (Lavoisier 1970). This volume reports experiments involving calcination, combustion, and dissolution of earths such as chalk that support the existence of an "elastic flexible fluid" in chalk, alkalis, and metallic calxes. He is not at all clear whether this fluid is part of air or air itself: he speaks of "an elastic fluid of a particular kind which is mixed with the air" (p. 340), but suspends judgment on the relation of this fluid to atmospheric air. In a memoir read to the Academy in 1775, included as an appendix in the 1776 English translation, he asserts "that the principle that is united to metals during their calcination, which increases their weight, and which constitutes them in the state of a

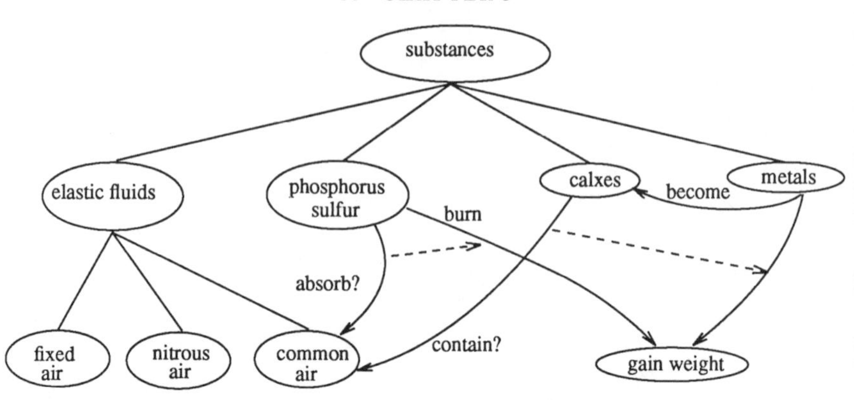

Figure 3.4. Fragment of Lavoisier's conceptual system in 1774. Straight lines indicate kind-relations. Curved lines with arrowheads indicate rules. The dotted lines indicate explanatory relations between rules.

calx, is neither one of the constituent parts of the air, nor a particular acid diffused in the atmosphere; that it is the air itself undivided" (p. 408). Thus in 1774 and early 1775 the structure of Lavoisier's conceptual system is roughly shown in Figure 3.4. Note that this is a small fragment of the structure displayed in the *Opuscules*, in which, for example, earths such as chalk receive much attention.

The major additions shown in Figure 3.4 over Figure 3.3 are the new results on combustion of phosphorus and sulfur. These substances gain weight in combustion, just as metals gain weight in calcination. Lavoisier now had some of the structure of the later oxygen theory: he explained combustion, calcination, and other phenomena by supposing the presence of air in minerals. But he still lacked a clear idea of what that air is. Moreover, he was not yet confident that he had a sharp alternative to the phlogiston theory of Stahl, saying of calcination and reduction that the present state of knowledge did not permit deciding between his and phlogiston interpretations and that the opinion of Stahl was perhaps not incompatible with his (Lavoisier 1970, 325). In 1776 Lavoisier admitted in correspondence that he often had more confidence in the ideas of the eminent British phlogiston theorist Joseph Priestley than in his own (Holmes 1985).

3.2.4 Lavoisier, 1777

By 1777, however, Lavoisier had developed a rich and clear alternative to the phlogiston theory. In several memoirs read to the Academy in that year, he described "pure air" or "eminently respirable air" as one of the ingredients of atmospheric air. Advances in experimental technique enabled Scheele, Priestley, and Lavoisier to isolate this ingredient. Priestley, true to the phlo-

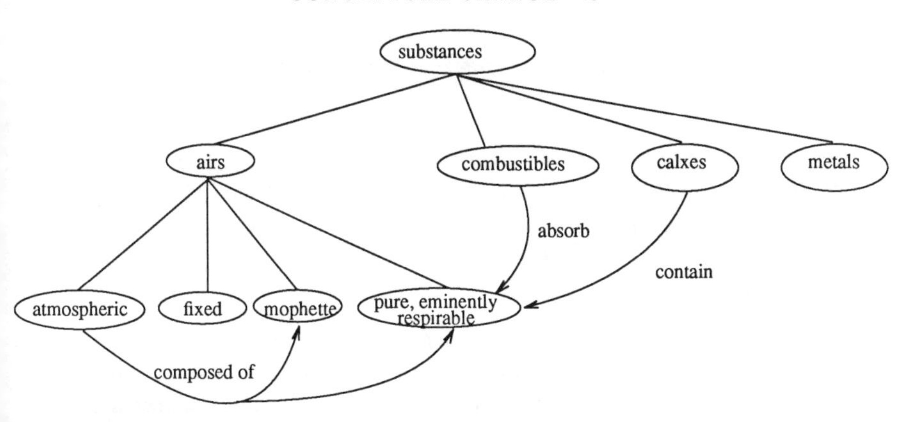

Figure 3.5. Fragment of Lavoisier's conceptual system in 1777. Straight lines indicate kind-relations. Curved lines with arrowheads indicate rules.

giston theory, called it "dephlogisticated air": combustion and respiration worked so well in it because its phlogiston had been removed. But Lavoisier was now convinced that the agent in combustion and calcination was a separate part of air. The final 1778 version of Lavoisier's memoir of 1775 (see Lavoisier 1970, Appendix) differs remarkably from the earlier version. The conclusion I quoted above from p. 408 becomes: "that the principle that unites with metals during their calcination, that increases their weight and constitutes them in the state of calxes, is nothing but the *portion of air the most salubrious and the most pure*" (Lavoisier 1862, vol. 2, 123, my translation and emphasis.)

Figure 3.5 depicts the relevant fragment of Lavoisier's conceptual system of 1777. His most systematic statement of his views at that time is in his "Mémoire sur la combustion en général" (Lavoisier 1862, 225–233; English translations can be found in Knickerbocker 1962, and in Leicester and Klickstein 1952). Lavoisier listed four kinds of air: eminently respirable, atmospheric, fixed, and mophette (nitrogen). Lavoisier described how combustion and calcination are subject to the same laws and how they can be given a common explanation by considering pure air as the real combustible body. (Respiration is discussed in another memoir of the same year.) Although he criticized the followers of Stahl for failure to isolate phlogiston, and suggested that the existence of an alternative hypothesis might shake the system of Stahl to its foundations, he did not feel sufficiently sure to reject the phlogiston theory out of hand. He concluded by saying: "In attacking here the doctrine of Stahl, it was not my purpose to substitute for it a rigorously demonstrated theory, but only an hypothesis which seemed to be more probable, more in conformity with the laws of nature, and one which appeared to involve less forced explanations and fewer contradictions" (Knickerbocker 1962, 134).

3.2.5 Lavoisier's mature theory: the 1780s

By 1783, however, the gloves were off, and Lavoisier demolished the doctrine of Stahl in his "Réflexions sur le Phlogistique" (Lavoisier 1862, 623–655). Using the term he coined in 1780, Lavoisier now referred to pure or eminently respirable air as *principe oxygine*. Lavoisier's position is presented in his opening paragraphs:

> In the series of memoirs that I have communicated to the Academy, I have reviewed the chief phenomena of chemistry; I have stressed those that accompany the combustion and calcination of metals, and, in general, all the operations where there is absorption and fixation of air. I have deduced all these explanations from a simple principle, that pure air, vital air, is composed of a particular principle belonging to it and forming its base, and that I have named *principe oxygine*, combined with the matter of fire and heat. Once this principle is admitted, the chief difficulties of chemistry seemed to fade and dissipate, and all the phenomena were explained with an astonishing simplicity. But if everything is explained in chemistry in a satisfactory manner without the aid of phlogiston, it is from that alone infinitely probable that this principle does not exist. (Lavoisier 1862, vol. 2, 623, my translation)

This quotation shows that Lavoisier had completely rejected the phlogiston theory, but his concept of oxygen clearly differs from ours in that oxygen gas was not itself an element. Rather, it was a compound of *principe oxygine* (as for Stahl, "principle" here means basic and original) and of the matter of fire and heat. By 1789 when Lavoisier published his textbook, *Traité Elémentaire de Chimie*, he referred to the matter of fire and heat as *caloric*. Just as the phlogiston theorists assumed that objects that burned must contain an inflammable principle, Lavoisier assumed from 1772 on that air must contain a principle of heat to explain why combustion produces heat. Air, rather than the combustible substances, was the source of heat. The substance caloric was accepted until the development of a kinetic theory of heat in the next century. Figure 3.6 shows part of the conceptual structure of Lavoisier's final system (Lavoisier 1789). Oxygen is now an element along with light, caloric, and the newly discovered gases hydrogen and nitrogen. Oxides are produced by the oxidation of metals, and nonmetallic objects burn in combination with oxygen to produce light and heat. Comparison of Figure 3.6 with Figures 3.2 and 3.3 shows that a substantial conceptual shift had occurred, restructuring the conceptual system of chemistry through extensive alterations of kind-relations. The next section shows that there were also important changes in part-relations.

To summarize briefly the development of Lavoisier's conceptual system, it is instructive to focus on two aspects: the development of the concept of

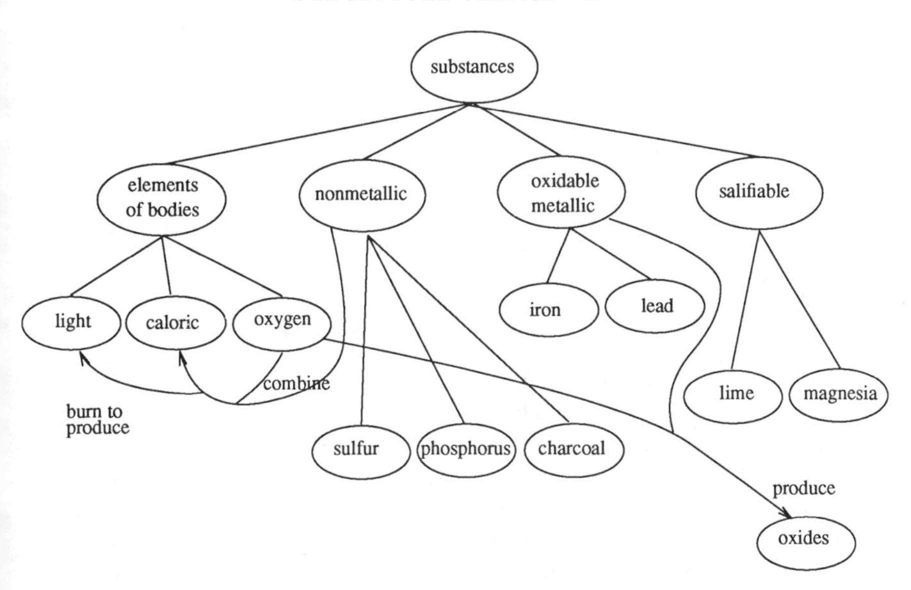

Figure 3.6. Fragment of Lavoisier's conceptual system in 1789. Straight lines indicate kind-relations. Curved lines with arrowheads indicate rules: oxygen and nonmetallic substances combine to produce caloric and light; oxygen and metallic substances produce oxides.

oxygen and the degree of confidence placed by Lavoisier in his whole system. In 1772, Lavoisier had only a vague idea that air could combine with metals. By 1774 he had much more evidence that this was true, but was still unclear whether it was air or a part of air that combined. By 1777 he knew that an eminently respirable portion of the air was responsible, and by the 1780s he had conceived it as an element constituting part of the atmosphere. Over the same years he went from some vague ideas about air relevant to calcination and combustion (1772, 1774) to a hypothesis that he clearly saw as a rival to the phlogiston theory (1777), to a fully worked out theory that obviated the phlogiston theory (1783). I shall now attempt to sketch a theory of conceptual change that can account for such transformations as these.

3.3 TOWARD A THEORY OF CONCEPTUAL CHANGE

A theory of conceptual change adequate to account for conceptual revolutions must be able to describe the mechanisms by which discoverers of new conceptual systems such as Lavoisier can build up their new systems by generating new nodes and links. These mechanisms must make possible the full range of kinds of conceptual change described in section 3.1. The theory must also explain how the new conceptual system can come to replace the old, as

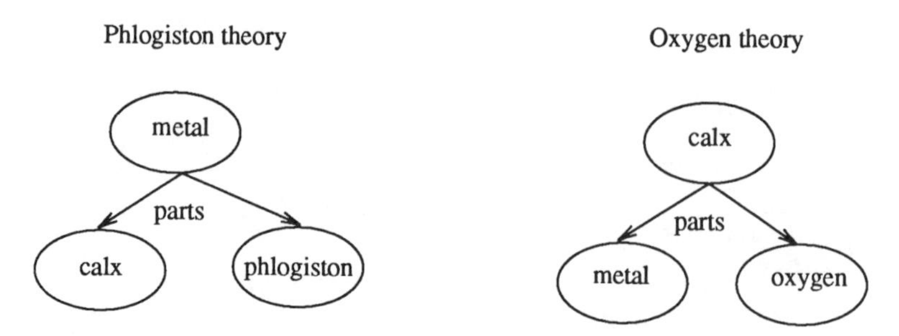

Figure 3.7. Part-relations in the phlogiston and oxygen theories.

the oxygen theory replaced the phlogiston theory. Finally, the theory must provide an account of how additional members of the scientific community can acquire and accept the newly constructed conceptual system.

Accounts of scientific change can be roughly divided into *accretion* theories and *gestalt* theories. On accretion views, a new conceptual system develops simply by adding new nodes and links. Such a theory would be inadequate for the case of Lavoisier, since it would neglect the enormous degree to which ideas were reorganized. Old concepts such as phlogiston were discarded and a whole new system of explanatory patterns was developed. In 1787 Lavoisier and his friends proposed a completely revised system of chemical nomenclature that substantially survives today. More than new terminology was involved, since old ideas were abandoned and a new conceptual organization was proposed. This is evident both in the great difference between the kind links in Figures 3.2 and 3.6, and in the difference between the part-whole views of Stahl and Lavoisier. Figure 3.7 contrasts the phlogiston theory's view of calxes and phlogiston as constituents of metals with the oxygen theory's view of metals and oxygen as constituents of calxes. Transformation from a phlogiston conceptual system to an oxygen conceptual system is much more complex than merely adding an oxygen node and deleting a phlogiston node, since other related nodes are affected too. The part-relations of the concepts of metal and calx change strikingly, going from a calx being a part of a metal to a metal being part of a calx. Another change in part-relations took place when water was decomposed into oxygen and hydrogen, overturning the view of Aristotle and Stahl that water is an element. Thus the chemical revolution involved reorganization of concepts as well as addition and replacement.

Kuhn (1970) persuasively criticized accretion theories of scientific growth. His "paradigms" were, among other things, radically different conceptual systems. Kuhn likened conceptual change to gestalt switches of the sort that occur in perceptual phenomena like the Necker cube in Figure 3.8, in which either face ABCD or face EFGH can be seen as the front. He used the fact that Priestley never accepted the oxygen theory as evidence for the incommensurability of paradigms.

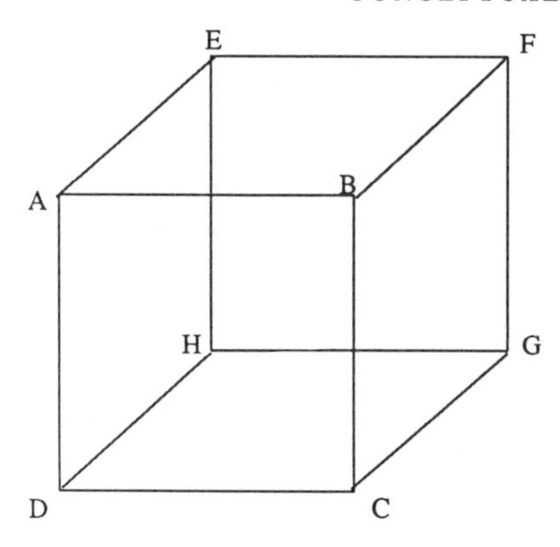

Figure 3.8. The Necker cube. Either face ABCD or EFGH can be seen as the front.

Thinking of changes as gestalt switches has the advantage over the accretion theory of taking seriously the degree of conceptual reorganization that goes into important scientific developments. But it makes it hard to see how conceptual change can take place. We saw that Lavoisier did not completely make the shift over to the new framework until around 1777. Holmes (1985) reports that even then Lavoisier caught himself in drafts of manuscripts using Priestley's term "dephlogisticated air," which he then deleted. Looking only at Stahl's conceptual system and Lavoisier's final one, it does appear that something like a gestalt switch has occurred. But understanding how Lavoisier got to his fully developed view requires consideration of the smaller steps that took him through the stages of 1772, 1774, and 1777.

Simple accretion or gestalt theories also have difficulty accounting for conceptual change in those who follow the discoverer in accepting a new conceptual framework. In the chemical revolution, radical conceptual change took place not only in Lavoisier, but in most chemists and physicists. By 1796 most scientists in Britain as well as in France had adopted the oxygen theory (Perrin 1988; McCann 1978). Priestley, who maintained the phlogiston theory until his death in 1804, was exceptional. Kuhn (1970) suggested, following a remark of Planck, that a revolutionary scientific view only triumphs because its opponents die. But Perrin's thorough chronicle shows that in the twenty years after 1775 virtually the entire scientific community converted over to Lavoisier's new system.

One of the last major proponents of the phlogiston theory was Richard Kirwan, who published a defense of the phlogiston theory in 1784. The French translation of Kirwan's *Essay on Phlogiston* was published in 1788 with responses by Lavoisier and his collaborators interspersed among its chapters (Kirwan 1968). These responses are fascinating, because they show Lavoisier and others systematically criticizing the attempts by Kirwan to defend the phlogiston theory. Rational dispute was clearly possible, and by 1792 Kirwan also had gone over to the oxygen theory. Radical conceptual change is not easy, but it is not impossible either.

We need a theory of conceptual change that avoids the weaknesses of both accretion and gestalt views and that accounts for revolutionary conceptual change in Lavoisier and his followers. Ideas from artificial intelligence can help greatly in constructing such a theory. I shall sketch computational accounts of both the *development* and *replacement* of conceptual systems, in both discoverers and those who acquire a new conceptual system from instruction by a discoverer. A full theory of conceptual change must account for four phenomena:

1. Development by discovery, in which someone sets up a new conceptual system, as Lavoisier did between 1772 and 1777.
2. Replacement by discovery, when the new conceptual system fully supplants the old, as happened to Lavoisier around 1777.
3. Development by instruction, when someone other than the discoverer becomes familiar with the new conceptual system by being told about it.
4. Replacement by instruction, when someone other than the discoverer adopts the new conceptual system and abandons the old.

To construct a fully general theory, I shall describe computational mechanisms that show how these conceptual changes can take place.

The mechanisms to be described are to be understood as part of the psychology of individuals, but it is easy to see how they can also have sociological consequences. A scientific revolution occurs only when a scientific community as a whole adopts a new conceptual system. The chemical revolution occurred because Lavoisier succeeded in constructing a new conceptual system that he passed on by instruction, first to his friends and eventually to the entire scientific community.

3.4 DEVELOPMENT OF CONCEPTUAL SYSTEMS BY DISCOVERY

How does a new conceptual system develop as Lavoisier's did from 1772 to 1789? In the early stages of this development, Lavoisier was undoubtedly capable of thinking within a phlogiston framework, for he sketches Stahlian explanations of some of the phenomena he was investigating (Lavoisier 1970). Thus he clearly had the phlogiston conceptual system and could apply it, even as he was developing the new conceptual framework of the oxygen theory using discoveries of his own and those of other researchers. We must consider how new concepts and the connections between them can be formed.

Fortunately, computational research on induction already suggests mechanisms for how this might work. The branch of artificial intelligence called *machine learning* offers a rapidly growing literature on concept and rule formation (see, for example, Michalski, Carbonell, and Mitchell 1983, 1986; Holland et al. 1986; Carbonell 1990). New concepts can be formed by a variety of bottom-up (data-driven) and top-down (theory-driven) methods. I shall not attempt here to survey the diversity of available methods, and for the sake of concreteness shall instead discuss mechanisms of the sort implemented in the artificial intelligence program PI (PI stands for "processes of induction" and is pronounced "pie." See Thagard and Holyoak 1985; Holland et al. 1986, ch. 4; Thagard 1988). PI implements the core idea of the framework laid out by Holland et al. (1986) that induction must be pragmatic, occurring in the context of problem solving, to assure relevance to the goals of the learner. There are an infinite number of trivial concepts that a learner might form, but for processing efficiency it is crucial not to clutter up the system with nodes like "blue-car-from-Japan-with-dented-bumper-and-New-Jersey-plates."

3.4.1 Conceptual Combination

In PI, new concepts are formed from old ones when combinations of features are shown to be relevant to the problem-solving operations of the system. For example, PI forms the theoretical concept of a sound wave after it postulates that sounds are waves to explain why sounds propagate and reflect. The conceptual combination *sound-wave* is interesting to the system because it is not simply the sum of sound and wave, since water waves, the source of the relevant concept of wave, move in single planes rather than in many planes like sound. PI's methods of concept formation show how new nodes similar to those that go into conceptual frameworks such as Lavoisier's can be formed in a pragmatically constrained fashion. By 1777 he had formed a new concept of "pure air" or "eminently respirable air" that later was relabeled "oxygen." This node obviously has great pragmatic value, since there are associated with it numerous highly predictive rules concerning the properties of such air: it supports combustion and respiration better than ordinary air, and it seems to Lavoisier to play a role in acids. Thus experiments conducted in 1775 and 1776 show the utility of a new concept of eminently respirable air differentiated from that of common air.

In Lavoisier's mature system, oxygen gas consists of oxygen principle plus caloric. Oxygen principle is a theoretical entity, not isolable in itself. In PI, such theoretical concepts can be formed by conceptual combination, as the example of sound wave shows. I can not now claim that Lavoisier's concept of oxygen developed using the mechanisms of the PI model, since substantiation of that claim would require still more detailed historical research and computational simulation in PI of the claimed developments. But it is at least clear how mechanisms like those in PI can contribute to the formation of new nodes such as oxygen. We can conjecture that Lavoisier combined the concepts *air* and *pure* because of the antecedent meanings of those concepts and the properties of the samples of gas to which the concept was applied: flames burned brighter and animals lived longer in them. Existing concepts and rules result in the formation of new concepts, which then produce the formation of new rules.

Recall the discussion of concepts as complex structures in section 2.5. If concepts are frame-like structures, then a new concept can be combined from old ones by creating new slots using the slots in the existing concepts. The PI system provided a simple method of adapting slots, but recent psychological research shows that more complex mechanisms are needed. Smith, Osherson, Rips, and Keane (1988) describe experiments that suggest that conceptual combination must take into account salience and diagnosticity. In their model, a concept *apple* has attributes of varying diagnosticity: color is more diagnostic for apples than shape or texture. Values of attributes, such as red and green for the color of an apple, vary in salience: for apples, red is more salient than green, since red apples are encountered more frequently than green ones. Using salience and diagnosticity, Smith et al. develop a model that appears adequate for simple cases of conceptual combination, but Medin and Shoben (1988) have shown experimentally that salience can vary with context. Small spoons are more typical (salient) than large spoons, but large wooden spoons are more typical than small wooden spoons. Moreover, Kunda, Miller, and Claire (1990) have shown that combination of social concepts is often influenced by causal reasoning that can produce emergent attributes not found in the original concepts. For example, subjects judge a blind lawyer to be determined, even though determination is not part of the original concepts *blind* and *lawyer*. Supposing that blind persons are determined helps to explain how they manage to become lawyers despite their handicap. Causal reasoning must also be important in combination of scientific concepts. A valuable research project would first identify complex kinds of scientific conceptual combination, and second develop a computational model sensitive to context effects and causal reasoning that could generate scientific concepts.

3.4.2 Generalization and Abduction

Given the presence of appropriate conceptual nodes, how are propositional links established between them? The links appropriate to the chemistry case appear to be general rules such as:

R1: Sulfur gains weight when it burns.
R2: Calxes contain air.

Although these two rules appear similar on the surface, very different mechanisms are required to form them. R1 is clearly a generalization from experimental observation, based on the experiments that Lavoisier performed in September of 1772. Lavoisier secured a sample of sulfur, did careful experiments with a powerful focusing lens, and concluded that sulfur that is burned gains in weight (Guerlac 1961). Generalization—the formation of general statements whose simplest form is "All A are B"—has been much investigated in the field of machine learning (see, for example, Langley et al. 1987). In PI, generalization takes into account not only the number of instances supporting a generalization, but also background knowledge of the variability of kinds of things involved. It appears that Lavoisier did not need many samples of sulfur to reach his generalization, since he expected from background knowledge about substances and combustion that the sample he had was typical of sulfur in general.

Generalization from experimental data is crucial for forming an important class of scientific rules, but it will not suffice for the formation of theoretical laws that go beyond what has been observed. Eighteenth-century chemistry—not to mention modern physics—had concepts intended to refer to a host of postulated entities. Lavoisier's stock included oxygen principle and caloric; his opponents' preeminent theoretical entity was of course phlogiston. In modern physics we have quarks and other postulated subatomic constituents of matter. Rules involving theoretical concepts that refer to nonobservable entities cannot be derived by empirical generalization since we have no observed instances from which to generalize. I am not assuming a strict distinction between theoretical entities and observable ones, since what counts as observable can change with technological advances such as the electron microscope; genes, which were originally theoretical entities, can now be observed. But in the early decades of the concept *gene*, rules about genes could not be formed by generalization from observations, since the properties of genes had to be postulated to explain observations.

The formation of rules that provide links with theoretical concepts requires a mechanism that I have called *rule abduction* (Thagard 1988, ch. 4). Peirce (1931–1958) coined the term "abduction" to refer to the formation of explanatory hypotheses. His general schema was:

Puzzling evidence E is to be explained.
Hypothesis H would explain E.
So maybe H.

This schema, however, does not look like a possible source of discoveries, since it appears that H has already been formed in the second premise. But abduction in PI has no such problem since, using a representation similar to that of predicate calculus for its rules, we can infer as follows:

Puzzling evidence **G(a)** is to be explained, i.e., why a is G.
Rule **(x)(Fx → Gx)**, i.e., all F are G, would explain **G(a)**.
So maybe **F(a)**, i.e., a might be F.

Such a schema can be applied to Lavoisier's early reasoning. Recall that in summer of 1772 his speculations that calxes contain air may have been based on two phenomena: effervescence and the gain of weight in calcination. For a particular piece of metal *m*, therefore, he can be imagined as setting out to explain why *m* effervesces in acid and gains weight during calcination. We then get the following abduction:

Explain: **effervesces(m).**
Rule: **contains(x, air) → effervesces(x).**
Abductive hypothesis: **contains(m, air).**

A similar kind of argument could use the rule that if something is added to a substance then it gains weight, along with additional information that the piece of metal was surrounded by air during calcination, to abduce that the piece of metal contains air after calcination. Rule abduction, then, is *generalization from abduced hypotheses* rather than from observed instances. From the doubly abduced hypothesis that *m* contains air PI can generalize that all instances of that kind of metal contain air. O'Rorke et al. (1990) describe a simulation of reasoning involving the increase in weight of mercury during calcination. Their program has a more sophisticated knowledge of physics principles than PI, but the basic mechanism of abductive hypothesis formation is similar to PI's.

We thus see in principle how conceptual links consisting of rules can be established by generalization from experiments and from the formation of theoretical hypotheses. The computational study of abduction is still in its infancy and much remains to be done to increase our understanding of how explanatory hypotheses are formed in science and ordinary life. One important question is the extent to which the formation of hypotheses is tied to their evaluation. The simplest model would be to have hypotheses formed according to minimal explanatory capability and then passed to an evaluator that would pick the best explanation. It might turn out, however, to be much more computationally efficient if the hypothesis generator worked closely with the evaluator. Chapter 4 describes how explanatory hypotheses can be evaluated.

3.4.3 Special Heuristics

The mechanisms so far discussed are general learning mechanisms applying to any domain where instances are found and facts are to be explained. The chemistry case suggests, however, that special heuristics might be useful for forming general rules. Langley et al. (1987, 228) propose the heuristic:

INFER-COMPONENTS
If A and B react to form C,
  or if C decomposes into A and B,
then infer that C is composed of A and B.

Specific examples of reasoning that appear to instantiate this heuristic can be found in Lavoisier (1970). However, Stahl's postulation of phlogiston, as well as the liquifiable and vitrifiable principles in earths, might be understood as the application of a different heuristic:

INFER-PRINCIPLE
If A has an important characteristic C,
then A contains a principle P that is responsible for C.

This heuristic has an unmodern ring, but Lavoisier seems to have used something like it himself in proposing caloric (the principle of heat) and oxygen principle.

The heuristic INFER-COMPONENTS lies at the center of the analysis of Langley et al. (1987) of the development of the phlogiston and oxygen theories (see also Zytkow and Simon 1986; Rose and Langley 1986). But it is important to recognize that it is too data-driven to account for the formation of those theories. It operates on such inputs as:

(reacts inputs {charcoal air} outputs {phlogiston ash air})
(reacts inputs {red-calx-of-mercury} outputs {mercury oxygen})

These are highly theoretical descriptions, since they include references to phlogiston and oxygen, which were theoretical entities not directly observed in any experiments. INFER-COMPONENTS therefore does not model the discovery of the phlogiston and oxygen theories, but only a later stage after the crucial concepts of phlogiston and oxygen have been formed. To form such concepts, more theory-based mechanisms such as conceptual combination, abduction, and the INFER-PRINCIPLE heuristic are needed. Abduction in particular is well described as explanation-driven, in contrast to data-driven generalizations. Not all discoveries, however, are data-driven or explanation-driven. In Chapter 8 we shall see cases where discoveries were coherence-driven, with new theories arising partly because of the need to overcome internal contradictions in existing theories.

Using discovery mechanisms such as concept formation, generalization, rule abduction, and special heuristics, new conceptual systems such as Lavoisier's can be developed. It is a separate question how such structures can replace competing ones.

3.5 REPLACEMENT BY DISCOVERY

How does a new conceptual system replace an existing one? The overthrow of the phlogiston theory was not accomplished by rejecting particular nodes or links, but by challenging the entire structure and replacing it with what by 1777 was a well-developed alternative. Accretion theories are fine for building up a new system: the discussion in the last section could be considered part of a computational accretion theory. But they will not account for how whole systems can be replaced.

Gestalt theories seem more plausible for replacement, since they allow that a whole system of relations can pop into place at once. But they have failed to specify how the new system can be constructed and how the replacement can occur. As a richer form of the gestalt metaphor, consider Figure 3.9. Stage (a) shows a conceptual system 1 with links to other concepts. Stage (b) shows a new conceptual system 2 partially formed in the background, also with links to other concepts. That the other concepts are linked to both systems runs counter to Kuhn's suggestion that in scientific revolutions the world changes. Although Priestley and Lavoisier had very different conceptual systems in 1777, there was an enormous amount on which they agreed concerning many experimental techniques and observations. Thus even revolutionary conceptual change occurs against a background of concepts that have relative stability. Finally in Figure 3.9, stage (c) shows system 2 fully developed and coming into the foreground, so that system 1 fades back. It does not disappear: Lavoisier could talk phlogiston-talk when required. The key question to be answered is: what makes system 2 pull into the foreground?

One nonholistic computational answer is based on rule competition. Holland (1986) describes rule systems in which rules have *strengths* that vary over time depending on the degree to which they have contributed to system performance. Similarly, PI assigns increased strength to rules that have figured significantly in problem solution. Many cognitive phenomena can be understood in terms of how rules with different strengths enter into *competition* with each other (Holland et al. 1986). In computational systems, even deduction must be pragmatically constrained to insure that what gets deduced may be relevant to the system's goals. Formal logic contains, for example, the rule of addition: from p to infer p *or* q. But any human or computer that used this deductively valid rule with abandon would quickly clog itself up with useless theorems. In deciding what rules to apply, utility is as important as probability, and rule strength captures aspects of both of these notions.

Let us suppose that the strengths of rules that provide links between concepts can be increased through successful use of those rules. In the scientific context, success consists mainly in providing explanations (on the nature of explanation, see section 5.3). When an explanation is accomplished, the links that made it possible can all be strengthened. Frequent use of a conceptual system in explanations will gradually build up the strength of all the rules in it to the point that they can become stronger than the rules in the existing system. It is at this point that stage (b) of Figure 3.9 gives way to stage (c), as the result of stronger rules emerging and dominating. This seems to be what happened in Lavoisier's case. In 1776 he was still sometimes thinking in phlogiston terms. But by 1777 the explanatory successes of his system based

(a) old conceptual network

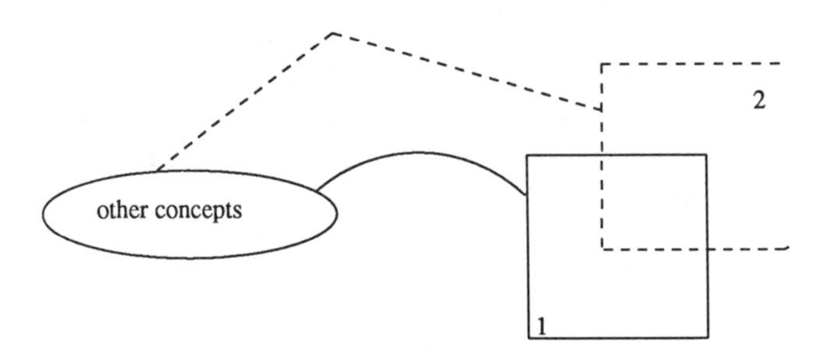

(b) partially constructed new network

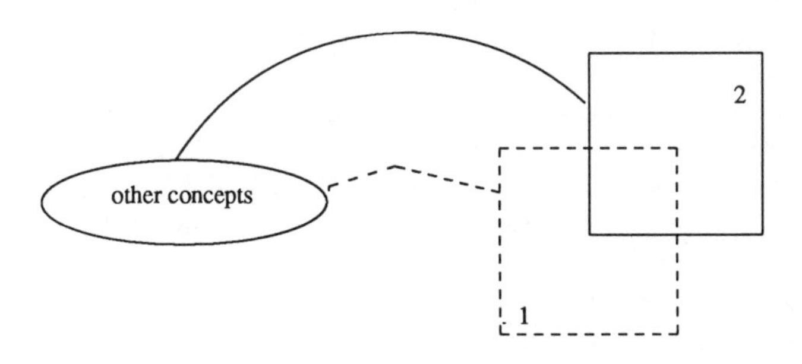

(c) new network supplants old

Figure 3.9. Conceptual change: a new network is built up and supplants an old one.

on eminently respirable air were so apparent that the links in his new system were stronger than the links in his old one. Thus replacement of one conceptual system by a newly developed one occurs if the new system gains enough problem-solving applications that its rules become stronger than those of the old system. This transformation may be an unconscious process occurring slowly. Old links are not deleted: they simply become weak enough that they no longer figure in the thought processes of the discoverer.

This rule-competition account of theory replacement has some plausibility, but it does not account for the holistic character of changes in conceptual systems. Scientists do not have to use individual rules over and over again to build them up, and the rules are not built up in isolation from each other. Rather, they can make dramatic shifts in conceptual systems when one set of hypotheses appears to them as a whole to have more *explanatory coherence* than another. Chapter 4 presents a theory of explanatory coherence that applies to the chemical and other revolutions, and Chapter 5 discusses the relations between successive theories.

3.6 DEVELOPMENT AND REPLACEMENT BY INSTRUCTION

Lavoisier's conceptual development required both the development of the oxygen theory and increase in its coherence to the point where it eclipsed the phlogiston theory. Similarly, scientists learning about the new framework by instruction need both to acquire and to strengthen it to the point where it can come to the fore.

Acquiring nodes and links by instruction is easier than acquiring them by discovery. New concepts can come by verbal communication: a phlogiston theorist could get "eminently respirable air" from the conversations or writings of Lavoisier or his friends. Rules that a discoverer would have to acquire by generalization or abduction can simply be communicated to another person who is spared the effort of initially forming them. Lavoisier, of course, did not himself discover all the empirical laws that fitted into his new system, but could by instruction acquire such crucial results as that metals gain weight in calcination.

Like discoverers, those acquiring a conceptual system by instruction have to incorporate and learn to think with the new system before it can be fully used. Education is not easy. Feeding concepts and rules into someone's head by rote is less important than developing this information in an organized way that can be applied. The inculcation of a conceptual system in scientists requires them to develop rules and procedures that are sufficiently coherent with other knowledge to supplant existing rules that otherwise would take precedence.

The major mechanism by which this strengthening takes place is scientific *argument*. It would be naive to suppose that arguments directly convince people. Rarely on an issue of any complexity and importance can you simply say to someone: Here are premises that you accept, from which the conclusion follows, so accept it. There are always responses to available arguments. But this does not mean that argument is futile, for the process of argument and later reflection on it can lead to revision of conceptual links, enabling an alternative system to come to the fore. The kinds of arguments that Lavoisier gave in favor of his theory in 1777 and 1783, as well as responses to opponents such as Kirwan, had great effect and led to general conversion. Arguments about the explanatory successes of the oxygen theory and the weaknesses of the phlogiston theory led most scientists of the day to accept the former. Development by instruction, setting up the requisite nodes and links, was not enough: people had to use the new system enough to appreciate its power. Chemists resistant to Lavoisier's ideas nevertheless repeated his experiments, thereby acquiring parts of his conceptual system. Perrin (1988) reports that it typically took several years for people to pass from opposition to Lavoisier's views to their acceptance. On my account, these years were spent both building up the new system and strengthening its links to where the new system seemed more coherent than the old. Conceptual reorganization of the kind-hierarchies and part-hierarchies took time.

The above shows how conceptual change is possible, but also makes it clear why it can be difficult. Lavoisier struggled for years to construct the conceptual edifice that became the oxygen theory. Not surprisingly, it sometimes took years for phlogistinians such as Kirwan to convert. Why did Priestley not make the conceptual shift? In 1796 Priestley published his *Considerations on the Doctrine of Phlogiston* in which he criticized the "antiphlogistic theory" for being based on only a few experiments that could be explained also by the phlogiston theory. Priestley was clearly well aware of the arguments of the oxygen theorists, for he (ironically?) dedicates his pamphlet to the surviving collaborators of Lavoisier who had written the responses to Kirwan! Priestley's arguments are rather weak, and show lack of awareness of some of the experiments that had been done, but they do not seem to differ in kind from the sorts of arguments concerning data and theories used by Lavoisier. In fact, Priestley wavered toward Lavoisier's views in the mid-1780s, but was brought back to the phlogiston position by James Watt. Like any other new theory, the oxygen theory did have internal problems that made it open to criticism. Priestley (1796) acknowledges that the weight of phlogiston has never been established, but points out that the same is true of Lavoisier's caloric.

My explanation, then, for why Priestley never became an oxygen theorist is twofold. First, as the preeminent phlogiston theorist, he had the most elaborate conceptual system for the phlogiston theory and, having used it more than others, the strongest appreciation of its coherence. Second, he never used the oxygen system enough to fully appreciate that it was more coherent than the phlogiston system. Kirwan, in contrast, through the exercise of arguing against the oxygen theorists, came to see that the system of explanations offered by the oxygen theory was more coherent than those offered by the view he had initially defended.

Other explanations of resistance to theory change are possible. Kunda (1987) has shown that motivation can lead people to resist conclusions that would make them unhappy. Coffee drinkers, for example, are less likely to accept evidence that caffeine causes disease. We have no evidence, however, that Priestley's reluctance to accept the oxygen theory was motivated by personal goals (see section 5.2.1 for further discussion of motivated inference).

Table 3.2 summarizes the mechanisms that I have postulated to explain conceptual changes in Lavoisier and in those who acquired the oxygen conceptual framework from him. Development by discovery requires mechanisms for concept formation, generalization, and hypothesis formation such as those being investigated in PI and other machine-learning programs. Replacement of a whole system of concepts and rules takes place by virtue of principles of explanatory coherence that can be implemented by an algorithm for selecting complexes of coherent hypotheses. Those who acquire a conceptual system from its discoverer must first through instruction and use build up an integrated set of concepts and rules, and second through argument come to see its explanatory coherence. In this way, a theory of explanatory coherence can be the driving force behind theory replacement and major conceptual change. The crucial point is that the new conceptual system does not arise by piecemeal modification of the old one. Rather, the new one must be built up largely on its own, and its replacement of the old is the result of a global judgment of explanatory coherence. Chapter 4 will describe the nature of such judgments.

Table 3.2
Mechanisms of Conceptual Change

	Discovery	*Instruction*
Development	concept formation	terms introduced
	generalization	experiments reported
	hypothesis formation	hypotheses defended
Replacement	use of new network shows its explanatory coherence	arguments with proponents of new network show its explanatory coherence

3.7 SUMMARY

Conceptual changes that involve alterations in kind-relations and part-relations are much more important to the development of scientific knowledge than most belief revision. The chemical revolution dramatically illustrates how the replacement of one theory by another can substantially alter conceptual hierarchies. The oxygen theory differed from the phlogiston theory in using different concepts and in organizing concepts in very different kind-hierarchies and part-hierarchies. The case study shows that conceptual change cannot be understood purely in terms of local accretions, but involves the replacement of entire conceptual systems. Such revolutionary conceptual changes are slow and difficult, but most scientists are capable of making the transition to a new theory and its attendant conceptual system.

CHAPTER 4

Explanatory Coherence

WHY did the oxygen theory of combustion supersede the phlogiston theory? Why is Darwin's theory of evolution by natural selection superior to creationism? This chapter develops a theory of explanatory coherence that applies to the evaluation of competing hypotheses in cases such as these. The theory is implemented in a connectionist computer program with many interesting properties. Both the theory and the program have been improved since their original presentation (Thagard 1989).

The theory of explanatory coherence, TEC for short, is central to the general theory of conceptual change in science. As we saw in the last chapter, conceptual revolutions require a mechanism that can lead people to abandon an old conceptual system and adopt a new one. The view of conceptual organization proposed in Chapters 2 and 3 renders implausible the claim that major conceptual replacements can take place in an incremental, evolutionary way. A more global mechanism that can install a new conceptual system is needed to account for such transitions, in which a new set of explanatory hypotheses intertwined with a new conceptual system replaces an old set. Whereas the last two chapters were concerned with how concepts are structured, this chapter describes how propositional systems are structured via relations of explanatory coherence.

The problem of inference to explanatory hypotheses has a long history in philosophy and a much shorter one in psychology and artificial intelligence. Scientists and philosophers have long considered the evaluation of theories on the basis of their explanatory power. In the late nineteenth century, C. S. Peirce discussed two forms of inference to explanatory hypotheses: *hypothesis*, which involved the acceptance of hypotheses, and *abduction*, which involved merely the initial formation of hypotheses (Peirce 1931–1958; Thagard 1988). Researchers in artificial intelligence and some philosophers have used the term "abduction" to refer to both the formation and the evaluation of hypotheses. AI work on this kind of inference has concerned such diverse topics as medical diagnosis (Pople 1977; Peng and Reggia 1990; Josephson et al. 1987) and natural language interpretation (Hobbs, Stickel, Appelt, and Martin 1990; Charniak and McDermott 1985). In philosophy, the acceptance of explanatory hypotheses is usually called *inference to the best explanation* (Harman 1973, 1986). In social psychology, attribution theory considers how people in everyday life form hypotheses to explain events (Fiske and Taylor 1984). Pennington and Hastie (1986, 1987) have proposed that much of jury decision making can be best understood in terms of explanatory coherence. For example, to gain a conviction of first-degree murder, the prosecution must convince the jury that the accused had a preformed intention to kill the victim. Pennington and Hastie argue that whether the jury will believe this depends on the explanatory coherence of the prosecution's story compared to the story presented by the defense.

Actual cases of scientific reasoning suggest a variety of factors that go into determining the explanatory coherence of a hypothesis. How much does the hypothesis explain? Are its explanations economical? Is the hypothesis similar to ones that explain similar phenomena? Is there an explanation of why the hypothesis might be true? In legal reasoning, the question of explaining the hypothesis usually concerns motive: if we are trying to explain the evidence by supposing that the accused murdered the victim, we will find the supposition more plausible if we can think of reasons why the accused was motivated to kill the victim. Finally, on all these dimensions, how does the hypothesis compare against alternative hypotheses?

This chapter presents a theory of explanatory coherence that is intended to account for a wide range of explanatory inferencces. I shall propose principles of explanatory coherence that encompass the considerations just described and that suffice to make judgments of explanatory coherence. Their sufficiency is shown by the implementation of the theory in a connectionist computer program called ECHO that has been applied to more than a dozen complex cases of scientific and legal reasoning. My account of explanatory coherence thus has three parts: the statement of a theory, the description of an algorithm, and the application to diverse examples that show the feasibility of the algorithm and help to demonstrate the power of the theory. Finally, I consider a series of objections to the theory, TEC, and to the implementation, ECHO. Chapter 5 describes in more detail how explanatory coherence contributes to conceptual change and discusses related issues concerning rationality and explanation.

4.1 A THEORY OF EXPLANATORY COHERENCE

Suppose that a professor is found dead, crushed beneath a Sun Workstation. Detectives investigating the case are assigned the task of determining what happened. Suppose further that a student was seen lurking around the professor's office, and that the student's fingerprints are found on the workstation. The detectives will then likely hypothesize that the student murdered the professor, since that hypothesis explains the death, the lurking, and the fingerprints. Of course, other hypotheses need to be taken into consideration: the murder might have been done by a dean, or it might have been an ingenious suicide; the fingerprints may be there because the student was doing some maintenance on the workstation. The detectives will naturally inquire whether the student had a motive, and they would be excited to find that the student had had a major quarrel with the professor about a course in which the student had been accused of submitting plagiarized work. The student's motive would explain the hypothesis that the student committed the murder, which then gains credibility from being explained as well as from explaining the evidence. Deciding whodunit is an exercise in explanatory coherence, requiring an assessment of the hypothesis that the student committed the murder on the basis of how well it fits with the evidence and other hypotheses. A theory of explanatory coherence should show how such assessments, in science as well as everyday life, can be made.

4.1.1 Coherence

Before presenting the theory, let me stress that I am not offering a general account of coherence. There are various notions of coherence in the literatures of different fields. We can distinguish at least the following:

Deductive coherence, which depends on relations of logical consistency and entailment among members of a set of propositions.
Probabilistic coherence, which depends on a set of propositions having probability assignments consistent with the axioms of probability.
Semantic coherence, which depends on propositions having similar meanings.

BonJour (1985) provides an interesting survey of philosophical ideas about coherence. Here, I am only offering a theory of *explanatory* coherence.

Explanatory coherence can be understood in several different ways, as

(a) a relation between two propositions,
(b) a property of a whole set of related propositions, or
(c) a property of a single proposition within a set of propositions.

I claim that (a) is fundamental, with (b) depending on (a), and (c) depending on (b). That is, explanatory coherence is primarily a relation between two propositions, but we can speak derivatively of the explanatory coherence of a set of propositions as determined by their pairwise coherence. Then we can speak derivatively of the explanatory coherence of a single proposition with respect to a set of propositions whose coherence has been established. A major requirement of an account of explanatory coherence is that it show how it is possible to move from (a) to (b) to (c). Algorithms for doing so are presented as part of the computational model described below.

Since the notion of the explanatory coherence of an individual proposition is so derivative and depends on a specification of the set of propositions with which it is supposed to cohere, I shall from now on avoid treating coherence as a property of individual propositions. Instead, we can speak of the *acceptability* of a proposition, which depends on but is detachable from the explanatory coherence of the set of propositions to which it belongs. We should accept propositions that are coherent with our other beliefs, reject propositions that are incoherent with our other beliefs, and be neutral toward propositions that are neither coherent nor incoherent.

In ordinary language, to cohere is to hold together, and explanatory coherence is holding together because of explanatory relations. We can accordingly start with a vague characterization:

Propositions P and Q cohere if there is some explanatory relation between them.

To fill this statement out we must specify what the explanatory relation might be. I see four possibilities:

1. P is part of the explanation of Q.
2. Q is part of the explanation of P.
3. P and Q are together part of the explanation of some R.
4. P and Q are analogous in the explanations they respectively give of some R and S.

This characterization leaves open the possibility that two propositions can cohere for nonexplanatory reasons: deductive, probabilistic, or semantic. Explanation is thus sufficient but not necessary for coherence. TEC takes "explanation" and "explain" as primitives (although see section 5.3), while asserting that a relation of explanatory coherence holds between P and Q if and only if one or more of (1)–(4) is true. *Incoherence* between two propositions occurs if they contradict each other or if they offer competing explanations.

4.1.2 Principles of Explanatory Coherence

I now propose seven principles that establish relations of explanatory coherence and make possible an assessment of the acceptability of propositions in an explanatory system S. S consists of propositions P, Q, and $P_1 \ldots P_n$. Local coherence is a relation between two propositions. I coin the term "incohere" to mean more than just that two propositions do not cohere: to incohere is to *resist* holding together. Here are the principles.

Principle 1. Symmetry.
(a) If P and Q cohere, then Q and P cohere.
(b) If P and Q incohere, then Q and P incohere.

Principle 2. Explanation.

If $P_1 \ldots P_m$ explain Q, then:

(a) For each P_i in $P_1 \ldots P_m$, P_i and Q cohere.

(b) For each P_i and P_j in $P_1 \ldots P_m$, P_i and P_j cohere.

(c) In (a) and (b) the degree of coherence is inversely proportional to the number of propositions $P_1 \ldots P_m$.

Principle 3. Analogy.[1]

If P_1 explains Q_1, P_2 explains Q_2, P_1 is analogous to P_2, and Q_1 is analogous to Q_2, then P_1 and P_2 cohere, and Q_1 and Q_2 cohere.

Principle 4. Data Priority.

Propositions that describe the results of observation have a degree of acceptability on their own.

Principle 5. Contradiction.

If P contradicts Q, then P and Q incohere.

Principle 6. Competition.

If P and Q both explain a proposition P_i, and if P and Q are not explanatorily connected, then P and Q incohere. Here P and Q are explanatorily connected if any of the following conditions holds:

(a) P is part of the explanation of Q,

(b) Q is part of the explanation of P,

(c) P and Q are together part of the explanation of some proposition P_j.

Principle 7. Acceptability.

(a) The acceptability of a proposition P in a system S depends on its coherence with the propositions in S.

(b) If many results of relevant experimental observations are unexplained, then the acceptability of a proposition P that explains only a few of them is reduced.

4.1.3 Discussion of the Principles

Principle 1, Symmetry, asserts that pairwise coherence and incoherence are symmetric relations, in keeping with the everyday sense of coherence as holding together. The coherence of two propositions is thus different from the nonsymmetric relations of entailment and conditional probability. Typically, P entails Q without Q entailing P, and the conditional probability of P given Q is different from the probability of Q given P. But if P and Q hold together, so do Q and P. The use of a symmetrical relation has advantages that

[1] In the original statement of TEC in Thagard (1989), Principle 3 included a second clause concerning disanalogies that is not included here because it lacks interesting scientific applications. The old Principle 7, system coherence, has similarly been deleted because it does little to illuminate actual scientific cases. A new Principle 6, competition, has been added to cover cases where noncontradictory hypotheses compete with each other. The old Principle 6, acceptability, becomes the new Principle 7.

will become clearer in the discussion of the connectionist implementation below.

Principle 2, Explanation, is by far the most important for assessing explanatory coherence, since it establishes most of the coherence relations. Part (a) is the most obvious: if a hypothesis P is part of the explanation of a piece of evidence E, then P and E cohere. Moreover, if a hypothesis P_2 is explained by another hypothesis P_1, then P_1 and P_2 cohere. Part (a) presupposes that explanation is a more restrictive relation than deductive implication, since otherwise we could prove that any two propositions cohere. Unless we use a relevance logic (Anderson and Belnap 1975), P_1 and the contradiction P_2 & not-P_2 imply any Q, so it would follow that P_1 coheres with Q. It follows from Principle 2(a), in conjunction with Principle 7, that the more a hypothesis explains, the more coherent and hence acceptable it is. Thus this principle subsumes the criterion of explanatory breadth (which William Whewell, 1967, called "consilience") that I have elsewhere claimed to be the most important for selecting the best explanation (Thagard 1978, 1988).

Whereas part (a) of Principle 2 says that what explains coheres with what is explained, part (b) states that two propositions cohere if together they provide an explanation. Behind part (b) is the Duhem-Quine idea that the evaluation of a hypothesis depends partly on the other hypotheses with which it furnishes explanations (Duhem 1954; Quine 1963). I call two hypotheses that are used together in an explanation "cohypotheses." Again I assume that explanation is more restrictive than implication, since otherwise it would follow that any proposition that explained something was coherent with every other proposition, because if P_1 implies Q, then so does P_1 & P_2. But any scientist who maintained at a conference that the theory of general relativity and today's baseball scores together explain the motion of planets would be laughed off the podium. Principle 2 is intended to apply to explanations and hypotheses actually proposed by scientists.

Part (c) of Principle 2 embodies the claim that if numerous propositions are needed to furnish an explanation, then the coherence of the explaining propositions with each other and with what is explained is thereby diminished. Scientists tend to be skeptical of hypotheses that require myriad ad hoc assumptions in their explanations. There is nothing wrong in principle in having explanations that draw on many assumptions, but we should prefer theories that generate explanations using a unified core of hypotheses. I have elsewhere contended that the notion of *simplicity* most appropriate for scientific theory choice is a comparative one preferring theories that make fewer special assumptions (Thagard 1978, 1988). Principles 2(b) and 2(c) together subsume this criterion. I shall not attempt further to characterize "degree of coherence" here, but the connectionist algorithm described below provides a natural interpretation. Many other notions of simplicity have been proposed (e.g. Harman et al. 1988; Foster and Martin 1966), but none is so directly relevant to considerations of explanatory coherence as the one embodied in Principle 2.

The third criterion for the best explanation in my earlier account was analogy, and this is subsumed in Principle 3. It is controversial whether analogy is of more than heuristic use, but scientists such as Charles Darwin have used analogies to defend their theories; his argument for evolution by natural selection is analyzed in Chapter 6. Principle 3 does not say simply that any two analogous propositions cohere. There must be an explanatory analogy, with two analogous propositions occurring in explanations of two other propositions that are analogous to each other. Recent computational models of analogical mapping and retrieval show how such correspondences can be noticed (Holyoak and Thagard 1989; Thagard et al. 1990).

Principle 4, Data Priority, stands much in need of elucidation and defense. In saying that a proposition describing the results of observation has a degree of acceptability on its own, I am not suggesting that it is indubitable, only that it can stand on its own more successfully than a hypothesis whose sole justification is what it explains. A proposition Q may have some independent acceptability and still not be accepted if it is only coherent with propositions that are not themselves acceptable.

From the point of view of explanatory coherence alone, we should not take propositions based on observation as independently acceptable without any explanatory relations to other propositions. As BonJour (1985) argues, the coherence of such propositions is nonexplanatory, based on background knowledge that observations of certain sorts are very likely to be true. From experience, we know that our observations are very likely to be true, so we should believe them unless there is substantial reason not to. Similarly, at a different level, we have some confidence in the reliability of descriptions of experimental results in carefully refereed scientific journals. Observations may be "theory-laden," as Hanson (1958) urged, but they are far from being theory-determined. I count as data not just individual observations such as "the instrument dial reads 0.5," but also generalizations from such observations. See section 9.1 for a discussion of the distinction between theoretical hypotheses and empirical generalizations.

Principle 5, Contradiction, is straightforward. By "contradictory" here I mean not just syntactic contradictions like P & not-P but also semantic contradictions such as "this ball is black all over" and "this ball is white all over." In my earlier version of TEC, I tried to stretch "contradiction" to cover cases where hypotheses that are not strictly contradictory are nevertheless held to be incompatible, but such cases are better handled by the new Principle 6, Competition.

According to Principle 6, we should assume that *hypotheses that explain the same evidence compete with each other unless there is reason to believe otherwise*. Hence there need be no special relation between two hypotheses for them to be incoherent, since hypotheses that explain a piece of evidence are judged to incohere unless there are reasons to think that they cohere. Not all alternative hypotheses incohere, however, since many phenomena have multiple causes. For example, explanations of why someone has certain medical symptoms may involve hypotheses that the patient has various diseases, and it is possible that more than one disease is present. Normally, however, if hypotheses are proposed to explain the same evidence, they will be treated as competitors. For example, in the debate over dinosaur extinction (Thagard 1991b), scientists generally treat as contradictory the hypotheses:

1. Dinosaurs became extinct because of a meteorite collision.
2. Dinosaurs became extinct because the sea level fell.

Logically, (1) and (2) could both be true, but scientists treat them as conflicting explanations. According to Principle 6, they incohere because both are claimed to explain why dinosaurs became extinct and there is no explanatory relation between them.

Principle 7, Acceptability, proposes in part 7(a) that we can make sense of the overall coherence of a proposition in an explanatory system just from the pairwise coherence relations established by principles 1–5. If we have a hypothesis P that coheres with evidence Q by virtue of explaining it, but incoheres with another contradictory hypothesis, should we accept P? To decide, we cannot merely count the number of propositions with which P coheres and incoheres, since the acceptability of P depends in part on the acceptability of those propositions themselves. We need a dynamic and parallel method of deriving general coherence from particular coherence relations; such a method is provided by the connectionist program described below.

Principle 7(b), reducing the acceptability of a hypothesis when much of the relevant evidence is unexplained by any hypothesis, is intended to handle cases where the best available hypothesis is still not very good in that it accounts for only a fraction of the available evidence. Consider, for example, a theory in economics that could explain the stock market crashes of 1929 and 1987 but had nothing to say about myriad other similar economic events. Even if the theory gave the best available account of the two crashes, we would not be willing to elevate it to an accepted part of general economic theory. What does "relevant" mean here? As a first approximation, we can say that a piece of evidence is *directly* relevant to a hypothesis if the evidence is explained by it or by one of its competitors. We can then add that a piece of evidence is relevant if it is directly relevant or if it is similar to evidence that is directly relevant, where similarity is a matter of dealing with phenomena of the same kind. Thus a theory of business cycles that applies to the stock market crashes of 1929 and 1987 should also have something to say about nineteenth-century crashes and major business downturns in the twentieth century.

According to TEC, a new theory will replace an old one if its hypotheses possess greater explanatory coherence. But TEC is still too vague to show how this could work. To show how to compute the acceptability of competing hypotheses, I now describe a program that implements TEC.

4.2 ECHO, A COMPUTATIONAL MODEL OF THEORY EVALUATION

4.2.1 Connectionist Models

To introduce connectionist techniques, I shall briefly describe the popular example of how a network can be used to understand the Necker cube phenomenon presented in section 3.3 (see, for example, Feldman and Ballard 1982; Rumelhart and McClelland 1986). Figure 3.8 contained a reversing cube: by changing our focus of attention, we are able to see as the front either face ABCD or face EFGH. The cube is perceived holistically, in that we are incapable of seeing corner A at the front without seeing corners B, C, and D at the front as well.

We can easily construct a simple network with the desired holistic property using *units*, crudely analogous to neurons, connected by links. Let Af be a unit that represents the hypothesis that corner A is at the front, while Ab represents the hypothesis that corner A is at the back. Similarly, we construct units Bf, Bb, Cf, Cb, Df, Db, Ef, Eb, Ff, Fb, Gf, Gb, Hf, and Hb. These units are not independent of each other. To signify that A cannot be both at the front and the back, we construct an *inhibitory* link between the units Af and Ab, with similar links inhibiting Bf and Bb, and so on. Because corners A, B, C, and D go together, we construct *excitatory* links between each pair of Af, Bf, Cf, and Df, and between each pair of Ab, Bb, Cb, and Db. Analogous inhibitory and excitatory links are then set up for E, F, G, and H. In addition, we need inhibitory links between Af and Ef, Bf and Ff, and so on. Part of the resulting network is depicted in Figure 4.1.

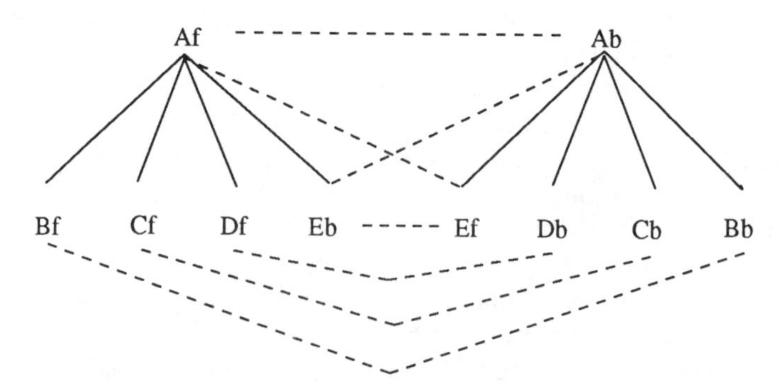

Figure 4.1. A connectionist network for interpreting the cube. Af is a unit representing the hypothesis that A is at the front, while Ab represents the hypothesis that A is at the back. Solid lines represent excitatory links, while dotted lines represent inhibitory links. Some inhibitory links are not shown.

Units can have varying degrees of *activation*. Suppose that our attention is focused on corner A, which we assume to be at the front, so that unit Af is activated. Then by virtue of the excitatory links from Af to Bf, Cf, and Df, these units will be activated. The inhibitory links from Af to Ab and Ef will cause those units to be deactivated. In turn, the excitatory links from Ab to Bb, Cb, and Db will deactivate them. Thus activation will spread through the network until all the units corresponding to the view that A, B, C, and D are at the front are activated, while all the units corresponding to the view that E, F, G, and H are at the front are deactivated.

4.2.2 ECHO, The Program[2]

Let us now look at ECHO, a computer program written in Common LISP that is a straightforward application of connectionist algorithms to the problem of explanatory coherence. In ECHO, propositions representing hypotheses and results of observation are represented by units. Whenever Principles 1–3 state that two propositions cohere, an excitatory link between them is established. If two propositions incohere, as specified by Principles 5–6, an inhibitory link between them is established. In ECHO, these links are symmetric, as Principle 1 suggests: the weight from unit 1 to unit 2 is the same as the weight from unit 2 to unit 1. Principle 2(c) says that the larger the number of propositions used in an explanation, the less the degree of coherence between each pair of propositions. ECHO therefore counts the propositions that do the explaining and proportionately lowers the weight of the excitatory links between units representing coherent propositions.

Principle 4, Data Priority, is implemented by links to each data unit from a special evidence unit that always has activation 1, giving each unit some acceptability on its own. When the network is run, activation spreads from the special unit to the data units, and then to the units representing explanatory hypotheses. The extent of data priority—the presumed acceptability of data propositions—depends on the weight of the link between the special unit and the data units. The higher this weight, the more immune the data units become from deactivation by other units. Units that have inhibitory links between them because they represent contradictory or competitive hypotheses have to vie with each other for the activation spreading from the data units: the activation of one of these units will tend to suppress the activation of the other. Excitatory links have positive weights, typically around .04 while inhibitory links have negative weights, typically around –.06. The activation of units ranges between 1 and –1; positive activation can be interpreted as acceptance

[2] Strictly speaking, this program is ECHO.2, an enhanced version that implements Principle 6, competition (Thagard 1991b). But I shall just call it ECHO.

of the proposition represented by the unit, negative activation as rejection, and activation close to 0 as neutrality.

To summarize how ECHO implements the principles of explanatory coherence, we can list key terms from the principles with the corresponding terms from ECHO.

Proposition: unit.
Coherence: excitatory link, with positive weight.
Incoherence: inhibitory link, with negative weight.
Data priority: excitatory link from special evidence unit.
Acceptability: activation.

Here are some examples of the LISP formulas that constitute ECHO's inputs. (I omit LISP quote symbols.)

1. (EXPLAIN (H1 H2) E1)
2. (EXPLAIN (H1 H2 H3) E2)
3. (EXPLAIN (H4) E1)
4. (ANALOGOUS (H5 H6) (E5 E6))
5. (DATA (E1 E2 E5 E6))
6. (CONTRADICT H1 H4)

Formula 1 says that hypotheses H1 and H2 together explain evidence E1. As suggested by the second principle of explanatory coherence proposed above, formula 1 sets up three excitatory links, between units representing H1 and E1, H2 and E1, and H1 and H2. Formula 2 sets up six such links, between each of the hypotheses and the evidence, and between each pair of hypotheses, but the weight on the links will be less than those established by formula 1, since there are more cohypotheses. Formula 3 sets up a single excitatory link between H4 and E1. In accord with Principle 3, Analogy, formula 4 produces excitatory links between H5 and H6, and between E5 and E6, if previous input has established that H5 explains E5 and H6 explains E6. Formula 5 is used to apply Principle 4, Data Priority, setting up explanation-independent excitatory links to each data unit from a special evidence unit. Formula 6 sets up an inhibitory link between the contradictory hypotheses H1 and H4, as prescribed by Principle 5. Finally, when all input has been read, ECHO implements Principle 6, Competition, by finding, for each piece of evidence E, pairs of hypotheses P and Q that explain E but are not explanatorily related to each other. Then an inhibitory link between P and Q is constructed. In the above example, since E1 is explained by H1 and H2 and by H4, ECHO sets up an inhibitory link between H2 and H4. No inhibitory link is set up between H1 and H2, since they are shown to be explanatorily related by formula 1; and no additional inhibitory link is needed between H1 and H4, since formula 6 already marked them as contradictory. In the limiting case, the inhibition established between competing hypotheses is the same as that between units representing contradictory hypotheses. But if P and Q each explain E only with the assistance of numerous other hypotheses, then they incohere to a lesser extent; compare Principle 2(c) above. Hence degree of inhibition between units representing P and Q is inversely proportional to the number of cohypotheses used by P and Q in their explanations of E, but proportional to the number of pieces of evidence E.

Once the input in formulas 1–6 and the principle of competition have created a network, ECHO uses a standard connectionist algorithm to adjust the activations of the units with respect to each other. A full specification of ECHO's inputs and algorithms is provided in the appendix at the end of this chapter.

Input to ECHO can optionally reflect the fact that not all data and explanations are of equal merit. For example, a data statement can have the form

(DATA (E1 (E2 .8))).

This formula sets up the standard link from the special unit to E1, but interprets the ".8" as indicating that E2 is not as reliable a piece of evidence as E1. Then the weight from the special unit to E2 is only 0.8 as strong as the weight from the special unit to E1. In medical cases, statistics exist to indicate the frequency of various diseases in the population. A proposition attributing a disease to someone is a hypothesis intended to explain various symptoms, but background statistical information can be used to treat the hypothesis as a weak form of data. For example, if there are two diseases D1 and D2 that both explain the same set of symptoms, but D1 is much rarer than D2, ECHO can be given such input as:

(DATA ((D1 .0001) (D2 .01))).

Then, other things being equal, ECHO will choose D2 over D1.

EXPLAIN statements can also take an optional numerical parameter, as in:

(EXPLAIN (H1) E1 .9).

The additional parameter ".9" indicates some weakness in the quality of the explanation and results in a lower than standard weight on the excitatory link between H1 and E1. In ECHO's applications to date, the additional parameters for data and explanation quality have not been used, since it is difficult to establish them objectively from the texts we have been using to generate ECHO's inputs. But it is important that ECHO has the capacity to make use of judgments of data and explanation quality when these are available.

Program runs show that the networks thus established have numerous desirable properties. Other things being equal, activation accrues to units corresponding to hypotheses that explain more, provide simpler explanations, and are analogous to other explanatory hypotheses. The considerations of explanatory breadth, simplicity, and analogy are smoothly integrated. The networks are holistic, in that the activation of every unit can potentially have an effect on every other unit linked to it by a path, however lengthy. Nevertheless, the activation of a unit is directly affected only by those units to which it is linked. Although complexes of coherent propositions are evaluated together, different hypotheses in a complex can finish with different activations, depending on their particular coherence relations. The symmetry of excitatory links means that active units tend to bring up the activation of units with which they are linked, while units whose activation sinks below 0 tend to bring down the activation of units to which they are linked. Data units are given priority, but can nevertheless be deactivated if they are linked to units that become deactivated. So long as excitation is not set too high (see section 4.6.2), the networks set up by ECHO are stable: in most of them, all units reach asymptotic activation levels after around 100 cycles of updating. To illustrate ECHO's capabilities, I shall describe some simple tests of how it handles considerations of explanatory breadth, simplicity, and analogy. Later in this chapter I present a more serious scientific example based on Lavoisier's argument against the phlogiston theory, and subsequent chapters include additional examples.

4.2.3 Explanatory Breadth

Other things being equal, we should prefer a hypothesis that explains more than alternative hypotheses. If hypothesis H1 explains two pieces of evidence while H2 explains only one, then H1 should be preferred to H2. Here are five formulas given together to ECHO as input:

(EXPLAIN (H1) E1)
(EXPLAIN (H1) E2)
(EXPLAIN (H2) E2)
(CONTRADICT H1 H2)
(DATA (E1 E2))

These generate the network pictured in Figure 4.2, with excitatory links corresponding to coherence represented by solid lines and inhibitory links corresponding to incoherence represented by dotted lines. By virtue of Principle 6, Competition, the same network would be generated if the CONTRADICT statement were omitted, because H1 and H2 compete to explain E2 and are not explanatorily related. Activation flows from the special unit, whose activation is clamped at 1, to the evidence units, and then to the hypothesis units, which inhibit each other. Since H1 explains more than its competitor H2, H1 becomes active, settling with activation above 0, while H2 is deactivated, settling with activation below 0. Notice that although the links in ECHO are symmetric, in keeping with the symmetry of the coherence relation, the flow of activation is not, since evidence units get activation first and then pass it along to what explains them.

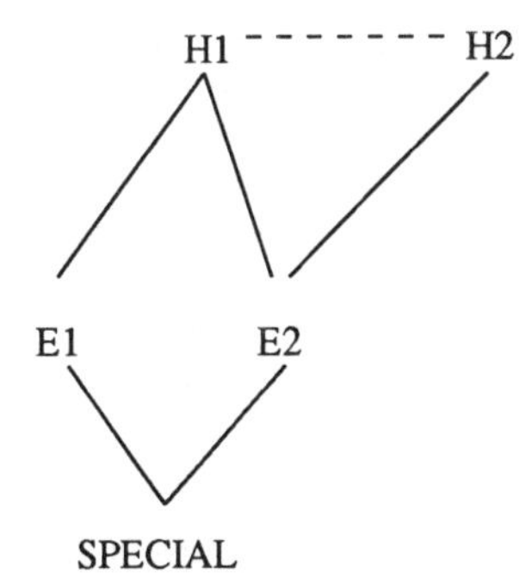

Figure 4.2. Explanatory breadth. As in figure 4.1, solid lines represent excitatory links, while dotted lines represent inhibitory links. Evidence units E1 and E2 are linked to the special unit. The result of running this network is that H1 defeats H2.

ECHO's networks have interesting dynamic properties. What happens if new data come in after the network has settled? When ECHO is given the further information that H2 explains additional data E3, E4, and E5, then the network resettles into a reversed state in which H2 is activated and H1 is deactivated. However, if the additional information is only that H2 explains E1, or only that H2 explains E3, then ECHO does not resettle into a state in which H1 and H2 get equal activation. (It does give H1 and H2 equal activation if the input says that they have equal explanatory power from the start.) Thus ECHO displays a kind of conservatism also seen in human scientists.

4.2.4 Being Explained

Section 4.2.3 showed how Principle 2(a) leads ECHO to prefer a hypothesis that explains more than its competitors. The same principle also implies greater coherence, other things being equal, for a hypothesis that is itself explained. Consider the input:

(EXPLAIN (H1) E1)
(EXPLAIN (H1) E2)
(EXPLAIN (H2) E1)
(EXPLAIN (H2) E2)
(EXPLAIN (H3) H1)
(CONTRADICT H1 H2)
(DATA (E1 E2))

Figure 4.3 depicts the network constructed using this input. Here, and in all subsequent figures, the special evidence unit is not shown. In Figure 4.3, H1 and H2 have the same explanatory breadth, but ECHO activates H1 and deactivates H2 because H1 is explained by H3. Other things being equal, ECHO gives more activation to a hypothesis that is explained than to one that is not explained. If the above formulas did not include a CONTRADICT statement,

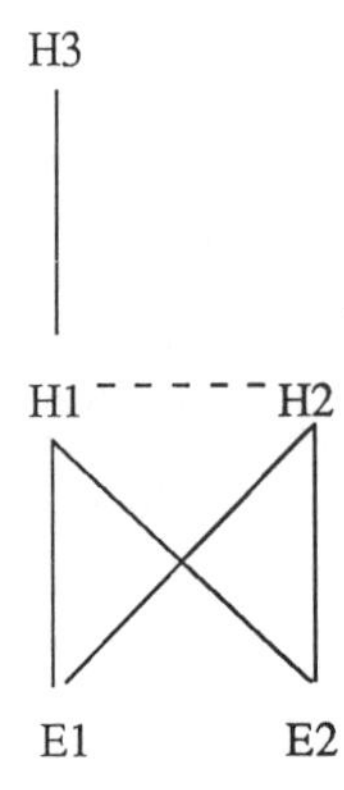

Figure 4.3. Being explained. H1 defeats H2 because it is explained by H3.

then no inhibitory links would be formed, so that all units would asymptote with positive activation. Because of the decay parameter, activation is still less than 1 (see the equations in section 4.6.1).

4.2.5 Refutation

According to Popper (1959), the hallmark of science is not the acceptance of explanatory theories but the rejection of falsified ones. Take the simplest case where a hypothesis H1 explains (predicts) some piece of "negative evidence" NE1 that contradicts data E1. Then E1 becomes active, deactivating NE1 and hence H1. Such straightforward refutations, however, are rare in science. Scientists do not typically give up a promising theory just because it has some empirical problems, and neither does ECHO. If in addition to explaining NE1, H1 explains some positive pieces of evidence E2 and E3, then ECHO does not deactivate it. However, an alternative hypothesis H2 that also explains E2 and E3 is preferred to H1, which loses because of NE1. Rejection in science is usually a complex process involving competing hypotheses, not a simple matter of falsification (Lakatos 1970; Thagard 1988, ch. 9).

4.2.6 Unification

The phenomena of explanatory breadth, being explained, and refutation all arise from Principle 2(a), which says that hypotheses cohere with what they explain. According to Principle 2(b), cohypotheses that explain together cohere with each other. Thus if H1 and H2 together explain evidence E, then H1 and H2 are linked. This gives ECHO a preference for unified explanations, ones that use a common set of hypotheses rather than having special hypotheses for each piece of evidence explained. Consider this input, which generates the network shown in Figure 4.4.

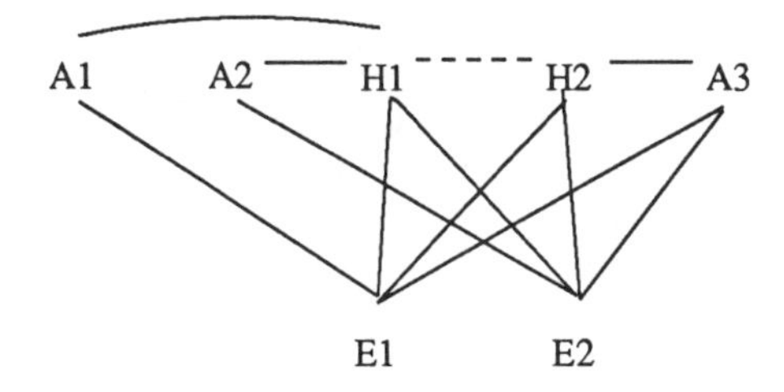

Figure 4.4. Unification. H2 defeats H1 because it gives a more unified explanation of the evidence.

(EXPLAIN (H1 A1) E1)
(EXPLAIN (H1 A2) E2)
(EXPLAIN (H2 A3) E1)
(EXPLAIN (H2 A3) E2)
(CONTRADICT H1 H2)
(DATA (E1 E2))

Although H1 and H2 both explain E1 and E2, the explanation by H2 is more unified in that it uses A3 in both cases. Hence ECHO forms a stronger link between H2 and A3 than it does between H1 and A1 or A2, so H2 becomes activated and H1 is deactivated. The explanations by H2 are not simpler than those by H1, in the sense of Principle 2(c), since both involve two hypotheses. ECHO's preference for H2 over H1 thus depends on the coherence of H2 with its auxiliary hypothesis and the evidence being greater than the coherence of H1 with its auxiliary hypotheses and the evidence.

4.2.7 Simplicity

According to Principle 2(c), the degree of coherence of a hypothesis with what it explains and with its cohypotheses is inversely proportional to the number of cohypotheses. An example of ECHO's preference for simple hypotheses derives from the input:

(EXPLAIN (H1) E1)
(EXPLAIN (H2 H3) E1)
(CONTRADICT H1 H2)
(DATA (E1))

Here H1 is preferred to H2 and H3 because it accomplishes the explanation with no cohypotheses. The generated network is shown in Figure 4.5.

Principle 2(c) is important for dealing with ad hoc hypotheses that are introduced only to save a hypothesis from refutation. Suppose that H4 is in danger of refutation because it explains negative evidence NE4 that contradicts evidence E4. One might try to save H4 by concocting an auxiliary hypothesis H5 that together with H4 would explain E4. But this ad hoc maneuver may not succeed in saving H4, especially if there are alternative hy-

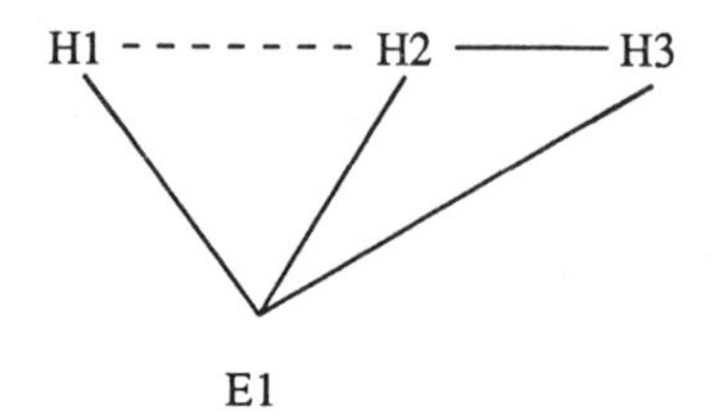

Figure 4.5. Simplicity. H1 defeats H2 because it gives a simpler explanation of the evidence.

potheses available. Because the explanation of E4 by H4 and H5 is less simple than the explanation of NE4 by H4, the activation that H4 receives from E4 may be less than the activation it loses because NE4 has a negative activation. Ad hoc maneuvers are common in science: nineteenth-century physicists did not abandon Newtonian mechanics because it gave false predictions about the motion of Uranus; instead, they hypothesized the existence of another planet, Neptune, to explain the discrepancies. Neptune, of course, was eventually observed, but we need to be able to discount auxiliary hypotheses that do not contribute to any additional explanations. Other things being equal, theories that employ auxiliary hypotheses will be less coherent than simpler ones. Note, however, that the simplicity of a theory is not merely a function of the number of hypotheses it has. In ECHO, the principle of simplicity operates locally, decreasing excitation on particular links. ECHO does not use a global count of the number of hypotheses in a theory to judge simplicity.

4.2.8 Analogy

According to Principle 3, analogous hypotheses that explain analogous evidence are coherent with each other. Figure 4.6 shows relations of analogy, derived from the input:

(EXPLAIN (H1) E1)
(EXPLAIN (H2) E1)
(EXPLAIN (H3) E3)
(ANALOGOUS (H2 H3) (E1 E3))
(CONTRADICT H1 H2)
(DATA (E1 E3))

The analogical links corresponding to the coherence relations required by Principle 3 are shown by wavy lines. Running this example leads to activation of H2 and deactivation of its rival H1. Figures 4.2–4.6 show explanatory breadth, simplicity, and analogy operating independently of each other, but in realistic examples these criteria can all operate simultaneously through activation adjustment. Thus ECHO shows how criteria such as explanatory breadth, simplicity, and analogy can be integrated. Principle 3 and ECHO

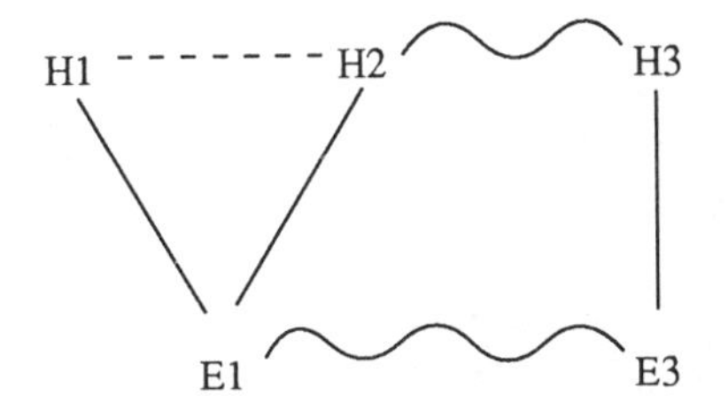

Figure 4.6. Analogy. The wavy lines indicate excitatory links based on analogies. H2 defeats H1 because the explanation it gives is analogous to the explanation afforded by H3.

allow analogy to participate with consilience and simplicity in contributing toward explanatory power.

4.2.9 Evidence and Acceptability

Principle 4 asserts that data get priority by virtue of their independent coherence. But it should be possible for a data unit nevertheless to be deactivated. We see this both in the everyday practice of experimenters, in which it is often necessary to throw out some of the data as unreliable (Hedges 1987), and in the history of science where evidence for a discarded theory sometimes falls into neglect (Laudan 1976). Figure 4.7, which derives from the following input, shows how this might happen.

(EXPLAIN (H1) E1)
(EXPLAIN (H1) E2)
(EXPLAIN (H1) E3)
(EXPLAIN (H1) E4)
(EXPLAIN (H2) E5)
(CONTRADICT H1 H2)
(CONTRADICT H1 E5)

These inputs lead to the deactivation of E5, which is dragged down by the deactivation of the inferior hypothesis H2 and inhibited by the superior hypothesis H1.

Principle 7(b) also concerns evidence, undermining the acceptability of hypotheses that explain only a small part of the relevant data. Accordingly,

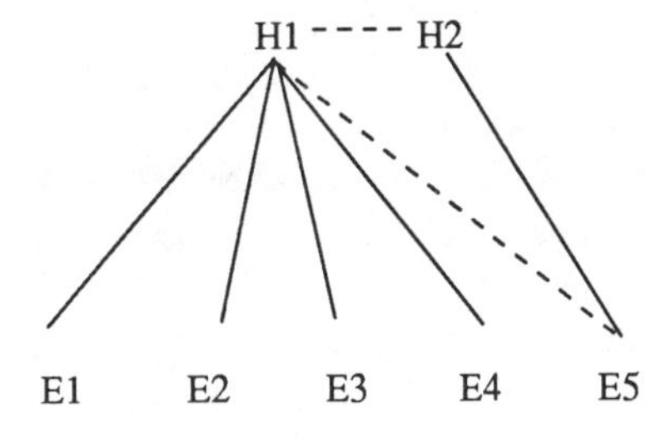

Figure 4.7. Downplaying of evidence. E5 is deactivated even though it is an evidence unit.

ECHO automatically increases the value of a decay parameter in proportion to the ratio of unexplained evidence to explained evidence (see section 4.6.1). A hypothesis that explains only a fraction of the relevant evidence will thus decay toward the beginning activation level of 0 rather than become activated.

If ECHO is taken as an algorithmic implementation of the first six principles of explanatory coherence, then it validates Principle 7, Acceptability, for it shows that holistic judgments of the acceptability of a proposition can be based solely on pairwise relations of coherence. A unit achieves a stable activation level merely by considering the activation of units to which it is linked and the weights on those links. Asymptotic activation values greater than 0 signify acceptance of the proposition represented by the unit, while negative values signify rejection.

4.2.10 Parameters

The simulations just described depend on program parameters that give ECHO numerous degrees of freedom, some of which are epistemologically interesting. In the example in Figure 4.2, the relation between excitatory weights and inhibitory weights is crucial. If inhibition is low compared to excitation, then ECHO will activate both H1 and H2, since the excitation that H2 gets from E1 will overcome the inhibition it gets from H1. Let the *tolerance* of the system be the absolute value of the ratio of excitatory weight to inhibitory weight. With high tolerance, the system will entertain competing hypotheses. With low tolerance, winning hypotheses deactivate the losers. Typically, the current version of ECHO is run with excitatory weights set at .04 and inhibition at –.06, so tolerance is 0.67. If tolerance is high, ECHO can settle into a state where two contradictory hypotheses are both activated. ECHO performs well using a wide range of parameters (see the sensitivity analyses in section 4.6.2). These particular parameter values were picked because they tend to lead to fast settling times, and the same values have been used in all simulations.

Other parameters establish the relative importance of simplicity and analogy. If H1 explains E1 by itself, then the excitatory link between H1 and E1 has the default weight .04. But if H1 and H2 together explain E1, then the weight of the links is set at the default value divided by 2, the number of cohypotheses, leaving it at .02. If we want to change the importance of simplicity as incorporated in Principle 2(c), however, then we can raise the number of cohypotheses to an exponent that represents the *simplicity impact* of the system. An equation for doing this is given in the algorithm section of section 4.6.1. The greater the simplicity impact, the more weights will be diminished by having more cohypotheses. Similarly, the weights established by analogy can be affected by a factor representing *analogy impact*. If this is 1, then the links connecting analogous hypotheses are just as strong as those set up by simple explanations, and analogy can have a large effect. If, on the other hand, analogy impact is set at 0, then analogy has no effect.

Another important parameter of the system is decay rate, represented by θ (see the equations in section 4.6.1). We can term this the *skepticism* of the system, since the higher it is, the more excitation from data will be needed to activate hypotheses. If skepticism is very high, then *no* hypotheses will be activated. Whereas tolerance reflects ECHO's view of contradictory hypotheses, skepticism determines its treatment of all hypotheses. Principle 7(b) can be interpreted as saying that if there is much unexplained evidence then ECHO's skepticism level is raised.

Finally, we can vary the priority of the data by adjusting the weights to the data units from the special unit. *Data excitation* is a value from 0 to 1 that provides these weights. To reflect the scientific practice of not treating all data equally seriously, it is also possible to set the weights and initial activations for each data unit separately. If data excitation is set low, then, contrary to section 4.2.3, new evidence for a rejected hypothesis will not lead to its adoption. If data excitation is high, then, contrary to section 4.2.9, evidence that supports only a bad hypothesis will not be thrown out.

With so many degrees of freedom, which are typical of connectionist models, one might question the value of simulations, since it might seem that any desired behavior whatsoever could be obtained. However, if a fixed set of default parameters applies to a large range of cases, then the arbitrariness is much diminished. In *all* the computer runs reported in this book, ECHO has had excitation at .04, inhibition at –.06, data excitation at .05, decay (skepticism) at .05, simplicity impact at 1, and analogy impact at 1. As reported in section 4.6.2 on sensitivity analyses, there is nothing special about the default values of the parameters: ECHO works over a wide range of values. In a full simulation of a scientist's cognitive processes, we could imagine better values being *learned*. Many connectionist models do not take weights as given, but instead adjust them as the result of experience. Similarly, we can imagine that part of a scientist's training consists in learning how seriously to take data, analogy, simplicity, and so on. Most scientists get their training not merely by reading and experimenting on their own, but by working closely with scientists already established in their field. Hence a scientist can pick up the relevant values from advisors. In ECHO they are set by the programmer, but it should be possible to extend the program to allow training from examples.

The examples described in this section are trivial, merely showing that ECHO has some desired properties. We shall see that ECHO can handle some much more substantial examples from the history of science.

4.3 APPLICATION OF ECHO TO THE CHEMICAL REVOLUTION

Theories in the philosophy of science, including computational ones, should be evaluated with respect to important cases from the history of science. This book will describe the application of ECHO to Lavoisier's argument for the oxygen theory, Darwin's argument for evolution by natural selection (Chapter 6), arguments for and against continental drift (Chapter 7), and the cases of Copernicus versus Ptolemy and Newton versus Descartes (Chapter 8). In addition, ECHO has been applied to:

- Contemporary debates about why the dinosaurs became extinct (Thagard 1991b).
- Jury decisions in murder trials (Thagard 1989).
- Adversarial problem solving, where one has to infer an opponent's intentions (Thagard 1991a).
- Psychological experiments on how beginning students learn physics (Ranney and Thagard 1988; Ranney 1991).
- Psychological investigations of how people perceive relationships (Miller and Read 1991).

Additional applications to reasoning in science, law, medicine, and education are currently under development. Let us now return to the major case of the last chapter, Lavoisier and the chemical revolution, and see how ECHO bears on the transition from phlogiston to oxygen.

As we saw in Chapter 3, the dominant chemical theory in the middle of the eighteenth century was the phlogiston theory of Stahl. The theory provided explanations of important phenomena of combustion, respiration, and calcination (what we would now call oxidation). According to the phlogiston theory, combustion takes place when phlogiston in burning bodies is given off. In the 1770s Lavoisier developed the alternative theory that combustion takes place when bodies combine rapidly with oxygen from the air. In 1783, more than ten years after he first suspected the inadequacy of the phlogiston theory, Lavoisier mounted a full-blown attack on it in a paper called "Réflexions sur le Phlogistique" (Lavoisier 1862).

Tables 4.1 and 4.2 present the input given to ECHO to represent Lavoisier's argument in his 1783 polemic against phlogiston. Table 4.1 shows the 8 propositions used to represent the evidence to be explained and the 12 used to represent the competing theories. The evidence concerns different properties of combustion and calcination, while there are two sets of hypotheses representing the oxygen and phlogiston theories, respectively. These propositions do not capture Lavoisier's argument completely but do recapitulate its major points. (In a more complicated simulation not presented here, I have encoded the attempt by the phlogiston theory to explain the increase in weight

Table 4.1
Input Propositions for Lavoisier Example

NOTE: Input to ECHO is stated in abbreviated form. The actual input for the first proposition is
(proposition 'E1 "In combustion, heat and light are given off.")

Evidence:

E1	In combustion, heat and light are given off.
E2	Inflammability is transmittable from one body to another.
E3	Combustion only occurs in the presence of pure air.
E4	Increase in weight of a burned body is exactly equal to weight of air absorbed.
E5	Metals undergo calcination.
E6	In calcination, bodies increase weight.
E7	In calcination, volume of air diminishes.
E8	In reduction, effervescence appears.

Oxygen hypotheses:

OH1	Pure air contains oxygen principle.
OH2	Pure air contains matter of fire and heat.
OH3	In combustion, oxygen from the air combines with the burning body.
OH4	Oxygen has weight.
OH5	In calcination, metals add oxygen to become calxes.
OH6	In reduction, oxygen is given off.

Phlogiston hypotheses:

PH1	Combustible bodies contain phlogiston.
PH2	Combustible bodies contain matter of heat.
PH3	In combustion, phlogiston is given off.
PH4	Phlogiston can pass from one body to another.
PH5	Metals contain phlogiston.
PH6	In calcination, phlogiston is given off.

in combustion and calcination by the supposition that phlogiston has negative weight. Lavoisier argues that this supposition renders the phlogiston theory internally contradictory, since phlogiston theorists sometimes assumed that phlogiston has positive weight.)

The propositions in Table 4.1 are part of Lavoisier's mature theory introduced in section 3.2.5. ECHO, however, deals only with relations between propositions, and does not take into account the structure of Lavoisier's conceptual system. It treats propositions as unanalyzed wholes, without breaking them down into their conceptual constituents. The descriptions of conceptual

Table 4.2
Input Explanations for Lavoisier Example

Oxygen explanations:	Phlogiston explanations:
(explain (OH1 OH2 OH3) E1)	(explain (PH1 PH2 PH3) E1)
(explain (OH1 OH3) E3)	(explain (PH1 PH3 PH4) E2)
(explain (OH1 OH3 OH4) E4)	(explain (PH5 PH6) E5)
(explain (OH1 OH5) E5)	
(explain (OH1 OH4 OH5) E6)	Data:
(explain (OH1 OH5) E7)	(data (E1 E2 E3 E4 E5 E6 E7 E8))
(explain (OH1 OH6) E8)	

organization in Chapter 3 had the opposite limitation: they largely omitted propositions and the explanatory relations among them. A full computational analysis of Lavoisier's mental system would specify all its conceptual and propositional components and clarify their interrelationships, but a cognitive system rich enough to do this and incorporate ECHO has yet to be built.[3]

Table 4.2 shows the part of the input that sets up the network used to judge explanatory coherence. The "explain" statements are based directly on Lavoisier's own assertions about what is explained by the phlogiston theory and the oxygen theory. In contrast to my earlier analysis of Lavoisier (Thagard 1989), I have not specified any contradictions between phlogiston hypotheses and oxygen hypotheses, since the hypothesis that burning objects give off phlogiston does not appear strictly to contradict the hypothesis that burning objects absorb oxygen. The current version of ECHO that implements TEC's principle of competition detects numerous competitions:

OH1 competes with PH5 because of E5.
OH1 competes with PH6 because of E5.
OH1 competes with PH3 because of E1.
OH1 competes with PH2 because of E1.
OH1 competes with PH1 because of E1.
OH2 competes with PH1 because of E1.
OH2 competes with PH2 because of E1.
OH2 competes with PH3 because of E1.
OH3 competes with PH3 because of E1.
OH3 competes with PH2 because of E1.
OH3 competes with PH1 because of E1.
OH5 competes with PH6 because of E5.
OH5 competes with PH5 because of E5.

[3] See Nelson, Thagard, and Hardy (1991) for a description of a new system that is a successor to PI (Thagard 1988) in integrating representations of concepts and rules, but makes extensive use of the mechanisms of parallel constraint satisfaction found in ECHO and in our analogy programs (Holyoak and Thagard 1989, Thagard et al. 1990).

These explanations and competitions generate the network portrayed in Figure 4.8. The link between OH1 and OH3 is particularly strong, since these two hypotheses participate in three explanations together. Figure 4.9, produced by a graphics program that runs with ECHO, displays the links to OH3, with excitatory links shown by thick lines and inhibitory links shown by thin lines. The numbers on the lines indicate the weights of the links rounded to three decimal places: in accord with Principle 2(c), weights for excitatory links are different from the default weight of .04 whenever multiple hypotheses are used in an explanation. If the hypotheses participate in only one explanation, then the weight between them is equal to the default excitation divided by the number of hypotheses; but weights are additive, so that the weight is increased if two hypotheses participate in more than one explanation. For example, the link between OH3 and E1 has the weight .013 (.01333333 rounded), since the explanation of E1 by OH3 required two additional hypotheses. The weight between OH3 and OH1 is .047 (.02 + .0133333 + .0133333), since the two of them alone explain E3, and together they explain E1 and E4 along with a third hypothesis in each case. OH1 and OH3 are thus highly coherent with each other by virtue of being used together in multiple explanations. The inhibitory links from OH3 to the three phlogiston hypotheses have the weight –.02, not the default inhibition –.06, because the relevant explanations each involve 3 cohypotheses (see section 4.6.1).

The numbers beneath the names in Figure 4.9 indicate the final activation of the named units, rounded to three decimal places. When ECHO runs this network, starting with all hypotheses at activation .01, it quickly favors the oxygen hypotheses, giving them activations greater than 0. In contrast, all the phlogiston hypotheses become deactivated. The activation history of the propositions is shown in Figure 4.10, which charts activation as a function of the number of cycles of updating. Figure 4.10 shows graphs, produced automatically during the run of the program, of the activations of all the units over the 106 cycles it takes them to reach asymptote. Notice that the oxygen hypotheses OH1–OH6 rise steadily to their asymptotic activations, while the phlogiston hypotheses sink to activation levels well below 0. PH4 does not directly compete with any oxygen hypotheses, but it is deactivated anyway since it is dragged down by the other phlogiston hypotheses. Thus the phlogiston theory fails as a whole.

This run of ECHO is biased toward the oxygen theory, since it was based on an analysis of Lavoisier's argument. We would get a different network if ECHO were used to model critics of Lavoisier such as Kirwan (1968), who defended a variant of the phlogiston theory. By the late 1790s, most chemists and physicists, including Kirwan, had accepted Lavoisier's arguments and rejected the phlogiston theory, a turnaround contrary to the suggestion of Kuhn (1970) that scientific revolutions occur when proponents of an old paradigm die off. The simulation presented in this chapter is intended to model,

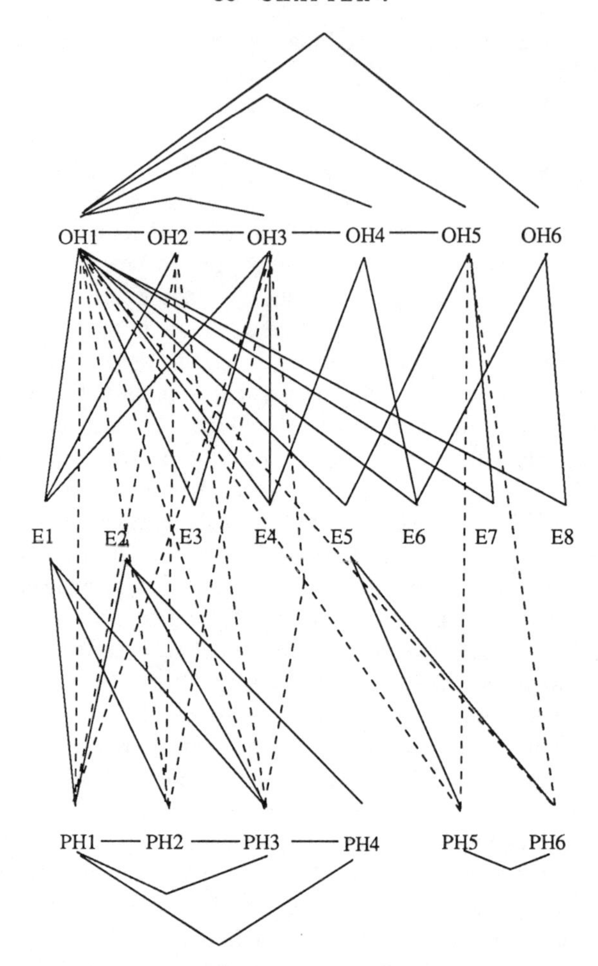

Figure 4.8. Network representing Lavoisier's argument. E1–E8 are evidence units. OH1–OH6 are units representing hypotheses of the oxygen theory, while PH1–PH6 represent the phlogiston hypotheses. Excitatory links, indicating that two propositions cohere, are represented by solid lines. Inhibitory links are represented by dotted lines.

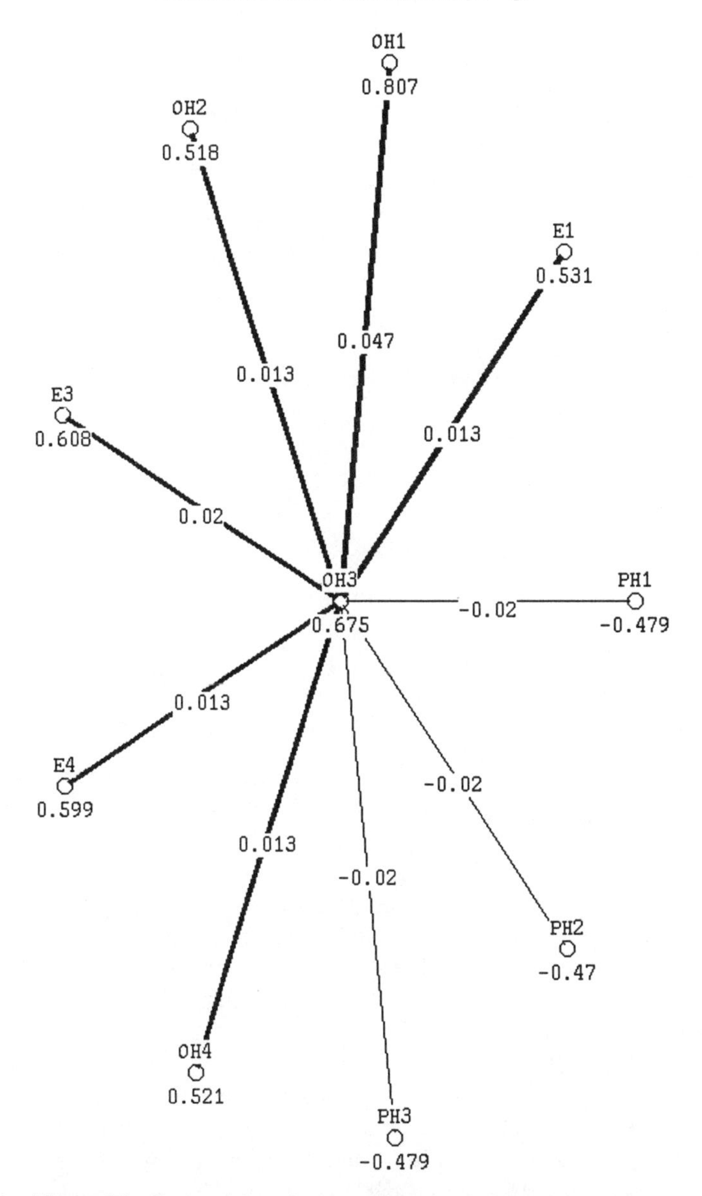

Figure 4.9. Connectivity of oxygen theory unit OH3. The numbers under the units are their activation values after the unit has settled. Thick lines indicate excitatory links, while thin lines indicate inhibitory links. Numbers on the lines indicate the weights on the links.

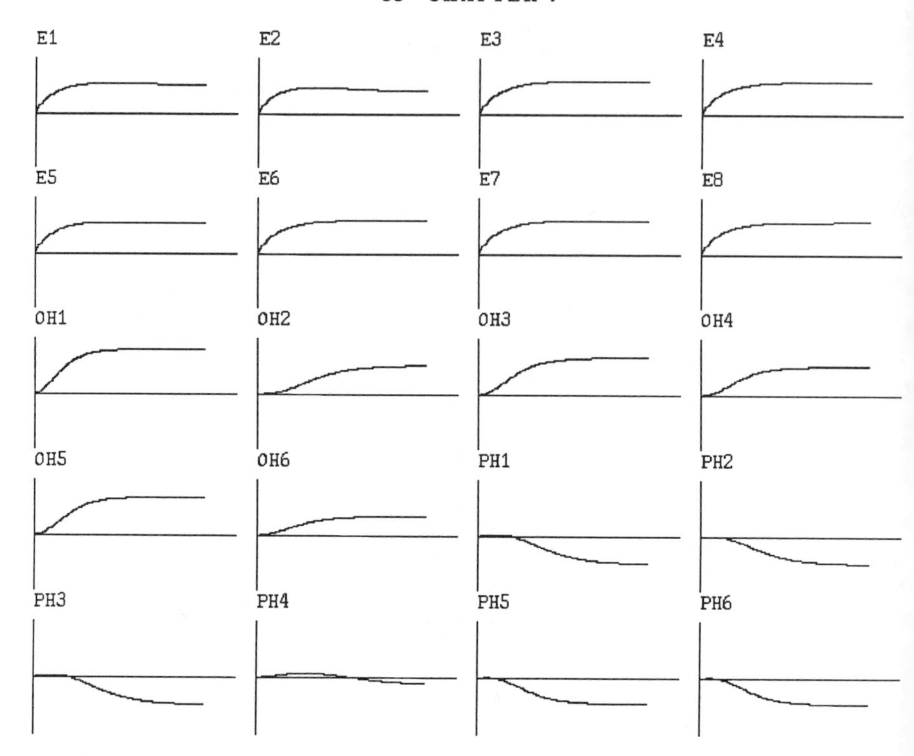

Figure 4.10. Activation history of Lavoisier network. Each graph shows the activation of a unit over 106 cycles of updating, on a scale of –1 to 1, with the horizontal line indicating the initial activation of 0.

from Lavoisier's own point of view, the acceptance of the oxygen theory and the rejection of phlogiston. It is not intended to represent the point of view of a phlogiston theorist, a neutral observer, or the entire scientific community. Nor does ECHO have anything to contribute to an understanding of how the oxygen theory was discovered.

Lavoisier's argument represents a relatively simple application of ECHO, showing two sets of hypotheses competing to explain the evidence. But more complex explanatory relations can also be important. Sometimes a hypothesis that explains the evidence is itself explained by another hypothesis. Depending on the warrant for the higher-level hypothesis, this extra explanatory layer can increase acceptability: a hypothesis gains from being explained as well as by explaining the evidence. The Lavoisier example does not exhibit this kind of coherence, since neither Lavoisier nor the phlogiston theorists attempted to explain their hypotheses using higher-level hypotheses, nor does the example display the role that analogy can play in explanatory coherence. Both these features are illustrated by the case of Darwin analyzed in Chapter 6.

4.4 OBJECTIONS AND REPLIES

Despite their many applications to important cases of scientific reasoning, the theory of explanatory coherence (TEC) and its computational implementation (ECHO) are highly controversial. Here I present a concise discussion of the objections most frequently made to TEC and ECHO. Many of these were made in different form by commentators on my 1989 *Behavioral and Brain Sciences* paper.

Objection 1. ECHO does little to illuminate the process of theory evaluation, since the simulations are all crafted by a programmer. The real work is in the encoding of ECHO input.

Reply. The charge that ECHO runs are contrived by the programmer has three replies, one historical, one psychological, and one computational. The historical reply is that every effort has been made to base input to ECHO's scientific examples on actual scientific texts. We have not made up examples to suit our purposes, but have tried to capture actual scientific arguments. The most detailed analyses so far, of Copernicus and Newton, have required more than 100 propositions each (Nowak and Thagard 1991a, 1991b). A logician might trivialize the analyses by conjoining all hypotheses in a theory into a single proposition, but such an exercise would be historically otiose.

The psychological reply to objection 1 is that ECHO encodings do not depend on any one programmer's idiosyncrasies, since numerous people have developed interesting ECHO applications. It is an open empirical question how general is the ability to analyze texts in ways that make sensible ECHO input, but Ranney (1991) and his students are actively investigating this question.

Finally, the computational reply is that ECHO can take its input not from human programmers, but from another program that generates the hypotheses that ECHO evaluates. A prototype of such a program is already running, but the full development will be in the context of a new cognitive system currently being built. Of course, a programmer will still have to encode the rules and other information that the hypothesis generator uses to provide input to ECHO, but the at least the ECHO input will not be directly fed to it. The development of this new system will also undercut the objection that ECHO is not really reasoning because it does not deal with the representation of propositions but treats them as unanalyzed wholes.

Objection 2. ECHO is unnecessarily complicated as an implementation of TEC. A simple linear function could compute the best explanation just as easily. Hobbs (1989) proposes a Naive Method: the score of a theory results from subtracting the number of hypotheses in a theory, #H, from the number of pieces of evidence it explains, #E. One theory should be accepted over another if it has higher value for #E–#H.

Reply. Hobbs' Naive Method (NM) will in fact work for many cases, since it captures the fundamental aspect of TEC that theories should explain more evidence with fewer assumptions. But NM is obviously not equivalent to ECHO. First, ECHO gives much more detailed output than NM, which evaluates theories—whole sets of hypotheses; ECHO generates an evaluation of individual hypotheses. NM accepts or rejects a set as a whole, while ECHO's output provides an acceptability value for each hypothesis. ECHO does not need to group hypotheses together into theories to evaluate them. It can reject one hypothesis that is part of an accepted theory, accept a hypothesis that is part of a rejected theory, and even reject pieces of evidence (4.2.9). ECHO automatically decides which hypotheses are in competition with each other. Thus ECHO is a more general mechanism of belief revision than NM.

Second, NM does not take into account the effect of being explained (4.2.4). Figure 4.3 presented a case where two hypotheses H1 and H3 are preferred to H2 alone, a clear violation of the NM's preference for lower #H. One might deny that a hypothesis becomes more acceptable if it is explained by another hypothesis. Yet in every domain in which explanatory inference is used, higher-level explanations can add to the acceptability of a hypothesis. Darwin, for example, thought that the fact that evolution was explained by natural selection was a crucial part of the evidence for evolution. In murder trials, questions of motive play a major role, since we naturally want an explanation of why the suspect committed the crime as well as evidence that is explained by the hypothesis that the suspect did it. In medical diagnosis, I expect that doctors normally find the hypothesis that cirrhosis of the liver is the cause of a patient's symptoms more convincing if they can also explain why the patient got cirrhosis from being a heavy drinker. ECHO shows that incorporating this element of explanatory coherence into a computational model does not create any intractable problems.

Third, unification (4.2.6) provides unlimited cases where ECHO exhibits a preference not found in NM. To take one of the simplest, consider a theory T1 consisting of hypotheses H1 and H2 that are both employed together to explain evidence E1 and E2. That is, H1 and H2 together explain E1, and together explain E2. The alternative explanations are H3 and H4, but H3 explains E1 alone and H4 explains E2 alone. Suppose that H1 contradicts H3 and H2 contradicts H4. T1 is more unified than the other singleton hypotheses, and ECHO indeed prefers H1 and H2, despite the fact that the Naive Method calculates #E – #H as 0 in both cases.

Fourth, NM does not take into account the effect of negative evidence and contradictions. Suppose H1 and H2 are competing explanations of E1, but H1 explains NE2 which contradicts evidence E2. #E – #H is the same, but in ECHO H2 wins. Similarly, NM does not take into account the effects of background knowledge. Suppose one theory explains more with fewer hypotheses than a rival theory, but contains hypotheses that contradict well-established hypotheses in another domain. For example, Velikovsky (1965) used the hypothesis that Venus passed near the Earth nearly 5,000 years ago to explain many historical events such as the reported parting of the Red Sea for Moses, but astronomers reject the hypothesis as inconsistent with Newton's laws of motion.

Fifth, NM does not take into account the effects of analogy, which ECHO naturally models by virtue of Principle 3 of TEC. There are thus numerous respects in which ECHO does a more interesting computation than NM. It is unlikely that any linear algorithm could compute what ECHO computes. The acceptability of a proposition depends not only on its coherences with other propositions, but on how acceptable those other propositions are. Contradictions and competition between propositions must be taken into account as well as coherences.

Anyone who likes TEC better than ECHO should appreciate that the relation between a theory and its computational implementation is not arbitrary. The organization of this chapter suggests that TEC came first and ECHO followed, but I got the idea for ECHO first, by analogy with the ACME program for analogical mapping that Keith Holyoak and I were developing (Holyoak and Thagard 1989). Thinking in terms of connectionist algorithms for simultaneously satisfying multiple constraints had enabled us to reconceptualize the problem of how the components of two analogs can be put in correspondence to each other, and it struck me that a similar approach might work for the problem of hypothesis evaluation. General ideas about inference to the best explanation and parallel constraint satisfaction led to ECHO, which led to TEC, and ECHO and TEC thereafter evolved together. As usual in cognitive science, there was considerable interplay of theory and model, with ideas about how to improve ECHO suggesting improvements in TEC and vice versa. The connectionist model thus played a crucial role in theory development, but it also has been instrumental in evaluating the theory. A typical theory in the philosophy of science is defended with brief discussion of a couple of examples. ECHO makes possible and necessary the development of detailed simulations that simultaneously lend credence to claims about the scope of ECHO and the scope of TEC. I therefore see connectionist ideas about parallel constraint satisfaction as integral to both the generation and the evaluation of a theory of explanatory coherence.

Objection 3. TEC and ECHO ignore the role of probabilities in scientific reasoning. A Bayesian theory of scientific theory choice renders them unnecessary.

Reply. My account of theory evaluation does indeed contrast with probabilistic accounts of confirmation that have been influential in philosophy since Carnap (1950) and are becoming increasingly important in AI (Pearl 1988).

Salmon (1966), for example, advocates the use of Bayes theorem for theory evaluation, which, if P(H/E) stands for the probability of H given E, can be written as:

$$P(H/E) = \frac{P(H)P(E/H)}{P(E)}$$

Consider what would be involved in trying to apply this to Lavoisier's argument against the phlogiston theory. We would have to take each hypothesis separately and calculate its probability given the evidence, but it is totally obscure how this could be done. Subjective probabilities understood as degrees of belief make sense in contexts where we can imagine people betting on expected outcomes, but scientific theory evaluation is not such a context. How could we take into account that alternative explanations are also being offered by the phlogiston theory? The issue is simplified somewhat if we consider only likelihood ratios for the oxygen and phlogiston theories, i.e., the ratio of P(E/oxygen) to P(E/phlogiston). But we still have the problem of dealing with the probability of the conjunction of a number of oxygen hypotheses whose degree of dependence is indeterminate.

Although Pearl's (1988) probabilistic approach appears promising for domains such as medical diagnosis where we can empirically obtain frequencies of co-occurrence of diseases and symptoms and thus generate reasonable conditional probabilities, it does not seem applicable to the cases of explanatory coherence I have been considering. What, for example, is the conditional probability of burned objects gaining in weight given the hypothesis that oxygen is combined with them? It would be 1 if the hypothesis entailed the evidence, but it does so only with the aid of the additional hypothesis that oxygen has weight and some unstated background assumption about conservation of weight. To calculate the conditional probability, then, we need to be able to calculate the conjunctive probability that oxygen has weight and that oxygen combines with burning objects, but these propositions are dependent to an unknown degree. Moreover, what is the probability that the evidence is true? In contrast to the difficulty of assigning probabilities to these propositions, the coherence relations established by my principles are easily seen directly in arguments used by scientists in their published writings. When frequencies are available because of empirical studies, probabilistic belief networks can be much more finely tuned than my coherence networks, but they are ill-suited for the kinds of nonstatistical theory evaluation that abounds in much of science and everyday life. The largest ECHO networks so far run would require thousands of values of conditional probabilities; even when the values are available, probabilistic networks can be computationally intractable (Pearl 1988; Peng and Reggia 1990; see Thagard 1991c for more extensive comparison of probabilistic and explanatory coherence approaches).

My recommendation is to use statistical inference and probabilistic reasoning whenever possible, but not to pretend it is always possible. It is *not* possible in theoretical contexts involving nonobservable entities, since such cases cannot provide the statistical information needed to get Bayesian reasoning going. In the example already discussed, we cannot get an estimate of the probability of something burning given that it combines with oxygen, since there is no way of determining the frequency of something combining with oxygen independent of the theory that postulates the existence of oxygen. To take a more mundane domain, we often explain other people's behavior in terms of intentions and desires that we attribute to them. To reduce such reasoning to Bayesian probability updating would require knowing the frequency with which people behave in a certain way given their intentions, but we have no independent way of establishing their intentions. Contrast this with van Fraassen's (1989) example of an inference that someone was eating based on observed dirty dishes. This case can be equally well described as an inference to the best explanation of why the dishes are dirty or as a probabilistic inference concerning the frequency of co-occurrence of dirty dishes and eating episodes. We can observe people eating, but we can only hypothesize people having intentions. Theoretical inferences, in science and ordinary life, require us to go beyond probabilistic reasoning based on frequencies. A philosophical response might be: so much the worse for theoretical inferences. That strategy would require us, however, to abandon completely the project of understanding scientific revolutions. But why trade Lavoisier, Darwin, Einstein, and other major episodes in the development of science for probabilistic purity?

Objection 4. TEC and ECHO assume that theories can be rationally compared, but philosophers of science such as Kuhn (1970) have shown that theories are incommensurable. TEC assumes basic agreement about what constitutes evidence and what constitutes explanation, but in scientific revolutions these become contentious issues.

Reply. We already saw in Chapter 3 that there was considerable conceptual overlap between the oxygen theorists and the phlogiston theorists. Later chapters will show that such overlap also existed in all other major cases of scientific revolutions. The important conceptual differences that undeniably exist in such cases are not so severe as to make ECHO analysis impossible. Oxygen and phlogiston proponents could recognize experiments done and explanations offered by each other, so incommensurability of the radical Kuhnian sort did not impede coherence-based adoption of the oxygen theory. In his recent writings, Kuhn discusses problems of incommensurabilty in terms of translation, a topic examined in section 5.2.3.

Objection 5. TEC ignores the fact that some hypotheses and pieces of evidence are more important than others, treating all explanations alike.

Reply. Importance is a slippery notion. TEC and ECHO do not contain any measure of the *practical* importance that some hypotheses might have. But a notion of theoretical importance emerges from using ECHO to analyze a scientific case. A hypothesis that coheres with numerous other hypotheses and pieces of evidence by virtue of its explanatory relations will be implicitly judged as much more important to a theory than a more isolated hypothesis. For example, in my analysis of Lavoisier's oxygen theory, OH3 is clearly much more important than OH6. Unlike Lakatos (1970), I am not proposing that theories have a rigid "hard core" of assumptions immune to revision. But from the perspective of TEC and ECHO it is easy to see how hypotheses that are intensely connected with other hypotheses in a theory will be harder to dislodge than isolated ones. As for pieces of evidence, their importance is a matter of their relation to alternative explanatory hypotheses. A data proposition will have greatest effect if it is explained by one hypothesis but not by its alternative.

To better understand how revising a belief can be affected by its relation to existing beliefs, we can define the *explanatory entrenchment* of a proposition as follows. Assume that previous judgments of explanatory coherence have been simulated by ECHO runs that identify some propositions as accepted (activation > 0) and others as rejected (activation < 0). Then the entrenchment of a proposition H could be:

(number of accepted propositions with which H coheres) – (number of accepted propositions with which H incoheres) – (number of rejected propositions with which H coheres) + (number of rejected propositions with which H incoheres)

Entrenchment does not itself predict whether a proposition will be accepted or rejected in a new run, since adding new information, such as new data and negative evidence, can lead previously accepted propositions to be rejected. But it provides an external way of understanding why some belief revisions are more difficult than others.

Objection 6. ECHO assumes a simplistic view of scientific theories as sets of propositions.

Reply. Theories are indeed more complicated than sets of propositions. A theory is a complex structure involving concepts, rules, and stored problem solutions (Thagard 1988, 40). But rules are the propositional parts of these structures: theories are not just sets of propositions, but they include the propositions that figure in explanatory relations.

Objection 7. ECHO is not really connectionist, since it does not use distributed representations.

Reply. This objection is partly terminological and partly substantive. The terminological aspect is handled by noticing, as we did in section 2.4, that there are different approaches to connectionist modeling. The term "connectionist" was used for local representations like ECHO's well before parallel distributed processing became the dominant form of connectionism (see Feldman and Ballard 1982; Rumelhart and McClelland 1986). The substantive criticism is that from a neurocomputational perspective, distributed representations are far more plausible and powerful (cf. Churchland 1989). But I argue in section 9.3.4, in the course of a general discussion of the place of connectionism in cognitive science, that local representations can be viewed as useful approximations to distributed representations.

Objection 8. TEC and ECHO say nothing about the nature of analogy, so Principle 3 and the input to ECHO using "ANALOGOUS" are left unspecified. Analogies are highly theoretical, so that there cannot be theory-independent input (compare objection 4).

Reply. Response to this objection would require a diversion from discussion of conceptual revolutions and presentation of a theory of analogy developed at length elsewhere (Holyoak and Thagard 1989; Thagard et al. 1990). If that theory is correct, then analogies can be recognized by satisfaction of a set of semantic, structural, and pragmatic constraints, without any overarching theory-dependent abstraction. In the general cognitive system now under development, ECHO will be able to receive its input concerning analogies from the analogy programs, without human intervention.

Objection 9. TEC concentrates on explanation, but scientists take much more seriously evidence that is predicted by a theory, not just explained after the fact. Hypotheses should lead to the prediction of new observations, not just manage the explanation of old ones (Popper 1959; Lakatos 1970).

Reply. Popperians have vastly exaggerated the importance of prediction. Lavoisier's case against the phlogiston theory was that his oxygen theory explained more; he says nothing about new predictions. Darwin's theory of evolution made little in the way of new predictions, but possessed much greater explanatory coherence than the hypothesis of divine creation (Chapter 6). A confirmed prediction was of some importance in the acceptance of plate tectonics (see the discussion of the Vine-Matthews hypothesis in section 7.3.3), but this was only one factor among several. Similarly, much has been made of the confirmation of Einstein's theory of relativity by the observation of the bending of light, but this prediction was no more important than the theory's explanation of puzzling facts already known such as the orbit of Mercury (see section 8.3). I have previously argued that the apparent preference for predictions over after-the-fact explanations is in fact a matter of simplicity Thagard (1988, 84). To explain something already known to have occurred, one can bring in all sorts of auxiliary hypotheses that diminish the simplicity of the explanation, but predictions are typically more pristine.

Popper valued predictions because they can lead to refutations, but his model rarely applies in the history of science. Typically, scientists do not react to a failed prediction by abandonment of their theories, but instead try to adapt and improve them. A Popperian scientist (if there were any) would be like a person who threw a car away because it did not start one morning. Scientists typically abandon a theory only when one with greater explanatory coherence comes along. Although ECHO does not exhibit simplistic Popperian falsification, it need not succumb to the various strategies that can be used to save a hypothesis from refutation. The strongest direct evidence *against* a hypothesis is pointing out that it has implications that contradict what has been observed. One way of saving the hypothesis from an objection of this sort is to use an auxiliary hypothesis to explain away the negative evidence. Section 4.2.7 showed how simplicity considerations can prevent this stratagem from working. Another way of saving a hypothesis in the face of negative evidence is to modify its cohypotheses. As Duhem (1954) and Quine (1963) pointed out, if H1 and H2 together imply some NE1 that contradicts a datum E1, then logic alone does not tell whether to reject H1, H2, or both. In ECHO, what hypotheses are deactivated depends on other relations of explanatory coherence. If H1 contributes to fewer explanations than H2, or if H1 contradicts another highly explanatory hypothesis H3, then H1 will be more likely to be deactivated than H2.

Objection 10. Accepting hypotheses in accord with TEC is a dangerous kind of inference. Inference to the best explanation is not merely evaluation of hypotheses with respect to how well they explain the evidence, since it selects only the best among the historically given hypotheses (van Fraassen 1989, 142). Unless we have some reason to believe that the truth is to be found in that set of hypotheses, we are not justified in believing hypotheses just because they possess explanatory coherence.

Reply. This objection raises difficult epistemological issues that I have discussed at length elsewhere (Thagard 1988, ch. 8). It would indeed be folly to recommend inference to the best explanation on the basis of the frequency with which it produces true beliefs, since there are many cases in the history of science where a theory that exceeds all others in explanatory coherence at a particular time is superseded. It was, for example, perfectly reasonable for a chemist before Lavoisier to accept the phlogiston theory, which provided a useful integrative explanation of various chemical phenomena. But science aims at more than the accumulation of true beliefs and the avoidance of false ones: understanding is as important a goal as truth, and the pursuit of it requires the acceptance of explanatory theories that may turn out to be false.

Objection 11. ECHO is intended as a cognitive model, but there is no reason to believe that it has psychological reality.

Reply. Any psychological claims about TEC and ECHO at this point have to be highly tentative, since there have only been limited applications to psychological experiments (Ranney and Thagard 1988; Ranney 1991; Miller and Read 1991). Several psychologists, however, are currently investigating ECHO and there should eventually be experimental evaluation of its strength as a cognitive model. Some commentators have wondered whether TEC and

ECHO are intended as descriptive of how people reason or prescriptive of how they should. I have coined a new term to describe an approach that is intended to be both descriptive and prescriptive (normative). I shall say that a model is "biscriptive" if it describes how people make inferences when they are in accord with the best practices compatible with their cognitive capacities. Unlike a purely prescriptive approach, a biscriptive approach is intimately related to actual human performance, rather than offering a theory of God's cognitive performance. But unlike a purely descriptive approach, biscriptive models can be used to criticize and improve human performance.

4.5 SUMMARY

TEC, the theory of explanatory coherence, provides a set of principles that establish relations of coherence and incoherence between propositions. The theory permits a smooth integration of diverse criteria such as explanatory breadth, simplicity, and analogy. ECHO's connectionist algorithm shows the computability of coherence relations. The success of the program derives from the usefulness of connectionist architectures for doing parallel constraint satisfaction and from the fact that inference to the best explanation is inherently a problem of simultaneously satisfying multiple constraints. ECHO has many properties found in scientists' reasoning, such as preferring simple and unified theories.

TEC captures the actual arguments of scientists such as Lavoisier who explicitly discuss what competing theories explain. There is no need to postulate probabilities or contrive deductive relations. TEC and ECHO engender a more detailed analysis of these arguments than is typically given by proponents of other accounts. Unlike most accounts of theory evaluation, this view based on explanatory coherence is inherently comparative. If two hypotheses contradict each other, they incohere, so the subsystems of propositions to which they belong will compete with each other. As ECHO shows, successful subsystems of hypotheses and evidence can emerge gracefully from local judgments of explanatory coherence. Thanks to ECHO, we know that there is an efficient algorithm for adjusting a system of propositions to turn coherence relations into judgments of acceptability. The algorithm allows every proposition to influence every other one, since there is typically a path of links between any two units, but the influences are set up systematically to reflect explanatory relations. Theory assessment is done as a whole, but a theory does not have to be rejected or accepted as a whole. Those hypotheses that participate in many explanations will be much more coherent with the evidence and with each other and will therefore be harder to reject. More peripheral hypotheses may be deactivated even if the rest of the theory they are linked to wins. We thus get a holistic account of inference that can nevertheless differentiate between strong and weak hypotheses. Although our hypotheses face evidence only as a corporate body, evidence and relations of explanatory coherence suffice to separate good hypotheses from bad.

4.6 APPENDIX: TECHNICAL DETAILS OF ECHO

For those interested in a more technical description of how ECHO works, this appendix outlines its principal algorithms and describes sensitivity analyses that have been done to determine the effects of the various parameters on ECHO's performance. LISP code is available from the author.

4.6.1 Algorithms

As described in section 4.2, ECHO takes as input EXPLAIN, CONTRADICT, ANALOGOUS, and DATA statements. The basic data structures in ECHO are LISP atoms that implement units with property lists that contain information about connections and the weights of the links between units. Table 4.3 describes the effects of the five kinds of input statements. All are straightforward, although the EXPLAIN statements require a calculation of the weights on the excitatory links. The equation for this is:

$$\textit{weight}(P, Q) = \textit{default weight} \,/\, (\textit{number of cohypotheses of } P)^{(\textit{simplicity impact})}$$

Here simplicity impact is an exponent, so that increasing it lowers the weight even more, putting a still greater penalty on the use of multiple assumptions in an explanation. In practice, however, I have not found any examples where it was interesting to set simplicity impact at a value other than 1.

After all input has been given, ECHO.2 automatically implements the principle of competition by finding for each proposition R pairs of hypotheses P and Q that explain R but are not explanatorily related to each other. Then an inhibitory link between P and Q is constructed. Table 4.4 gives the algorithm. The equation in part 3 of Table 4.4 ensures that in the limiting case where P and Q each explain R without any help, and R is the only propositon that they both explain, then the inhibition is the same as that between units representing contradictory hypotheses. The equation captures the following intuitions: the more propositions that P and Q both explain, the more they compete, but the greater the number of the additional assumptions used in these explanations, the less they compete. If P and Q each independently explains R only with the assistance of numerous other hypotheses, then they incohere to a lesser extent; compare Principle 2 (c), according to which the degree of *coherence* is lessened by the number of hypotheses that *together* explain a proposition. By virtue of the algorithm in Table 4.4, ECHO.2 tends to create many more in-

Table 4.3
Algorithms for Processing Input to ECHO

1. Input: (PROPOSITION NAME SENTENCE)
   Create a unit called NAME.
   Store SENTENCE with NAME.
2. Input: (EXPLAIN LIST-OF-PROPOSITIONS PROPOSITION)
   Make excitatory links between each member of LIST-OF-PROPOSITIONS and PROPOSITION.
   Make excitatory links between each pair of LIST-OF-PROPOSITIONS.
   Record what explains what.
   Note: weights are additive, so that if more than one EXPLAIN statement creates a link between two proposition units, then the weight on the link is the sum of the weights suggested by both statements. The weights are inversely proportional to the length of LIST-OF-PROPOSITIONS, implementing a principle of simplicity. Optionally, the EXPLAIN statement can end with a number representing the strength of the explanation; the weight on the link is then multiplied by this number.
3. Input: (CONTRADICT PROPOSITION-1 PROPOSITION-2).
   Make an inhibitory link between PROPOSITION-1 and PROPOSITION-2.
4. Input: (ANALOGOUS (PROPOSITION-1 PROPOSITION-2) (PROPOSITION-3 PROPOSITION-4))
   If PROPOSITION-1 explains PROPOSITION-3, and PROPOSITION-2 explains PROPOSITION-4, then make excitatory links between PROPOSITION-1 and PROPOSITION-2, and between PROPOSITION-3 and PROPOSITION-4.
5. Input: (DATA LIST-OF-PROPOSITIONS)
   For each member of LIST-OF-PROPOSITIONS, create an excitatory link from the special evidence unit with the weight equal to the data excitation parameter, unless the member is itself a list of the form (PROPOSITION WEIGHT). In this case the weight of the excitatory link to PROPOSITION is WEIGHT times the parameter.

hibitory links than did the original ECHO.1, which created them only when told that two propositions are contradictory.

After input has been used to set up the network, the network is run in cycles that synchronously update all the units. The algorithm for this is shown in Table 4.5. For each unit j, the activation a_j, ranging from –1 to 1, is a continuous function of the activation of all the units linked to it, with each unit's contribution depending on the weight w_{ij} of the link from unit i to unit j. The activation of a unit j is updated using the following equation.

$$a_j(t+1) = a_j(t)(1-\theta) + \begin{cases} net_j(max - a_j(t)) & \text{if } net_j > 0 \\ net_j(a_j(t) - min) & otherwise \end{cases}$$

Table 4.4
Algorithm for Implementing Principle 6, Competition

1. Compile a list of pairs of potentially competing hypotheses:
   (a) For each proposition R, compile a list of all the hypotheses that explain R. Create a list of pairs of the explainers of R.
   (b) Return a list L of pairs of potentially competing hypotheses.
2. Prune the list L by eliminating any pair (P Q) such that any of the following conditions holds:
   (a) P is in the list of explainers of Q,
   (b) Q is in the list of explainers of P,
   (c) P and Q are cohypotheses,
3. For each pair (P Q) in the pruned list L, create an inhibitory link between P and Q with a weight equal to:

$$\frac{\textit{inhib} * (\textit{number of propositions explained by both P and Q})}{(\textit{number of cohypotheses of P and Q in the explanations of propositions they both explain}) / 2}$$

Notes:
(i) *inhib* is a constant representing the default inhibition.
(ii) A unit u in ECHO is represented by a LISP atom with a property list that includes entries for units representing propositions that (1) explain, (2) are explained by, and (3) are cohypotheses of the proposition represented by U.
(iii) H1 and H2 are *cohypotheses* if H1 and H2 are together part of the explanation of some other proposition. In the calculation in 3, each proposition counts as a cohypothesis of itself.

Here θ is a decay parameter that decrements each unit at every cycle, *min* is minimum activation (–1), *max* is maximum activation (1), and net_j is the net input to a unit. This is defined by:

$$net_j = \sum_i w_{ij} a_i(t)$$

Repeated updating cycles result in some units becoming activated (getting activation > 0) while others become deactivated (activation < 0).

4.6.2 Sensitivity

The four most important parameters in ECHO are:

Excitation, the default weight of the link between units representing propositions that cohere, reduced in accord with Principle 2(c) if more than one hypothesis is involved in an explanation;

Table 4.5
Algorithms for Network Operation

1. Running the network:
   Set all unit activations to an initial starting value (typically .01), except that the special evidence unit is clamped at 1.
   Update activations in accordance with (2) below.
   If no unit has changed activation more than a specified amount (usually .001), or if a specified number of cycles of updating have occurred, then stop.
   Print out the activation values of all units.
2. Synchronous activation updating at each cycle:
   For each unit *j*, calculate the new activation of *j*, a_j, in accord with the equations below, considering the old activation a_i of each unit linked to *j*.
   Set the activation of *j* to the new activation.

Equations:

$$a_j(t+1) = a_j(t)(1-\theta) + \begin{cases} net_j(max-a_j(t)) & \text{if } net_j>0 \\ net_j(a_j(t)-min) & otherwise \end{cases}$$

Here θ is a decay parameter that decrements each unit at every cycle, *min* is minimum activation (–1), *max* is maximum activation (1), and net_j is the net input to a unit. This is defined by:

$$net_j = \sum_i w_{ij} a_i(t)$$

Inhibition, the default weight of the link between units representing propositions that incohere;

Decay, the amount that the activation of each unit is decremented on each cycle of updating; and

Data excitation, the weight of the link from the special evidence unit (whose activation is always 1) to each of the units representing pieces of evidence.

For each of the ECHO simulations mentioned in this book, hundreds of runs have been done to determine how different values of these parameters affect ECHO's performance. Experiments have determined that the last two parameters have little effect on simulations: greater decay values tend to compress asymptotic activation values toward 0, and greater data excitation tends to make the activations of losing units higher, but there are no qualitative differences. In contrast, the relative values of excitation and inhibition can be crucial. There are two ways in which a simulation can fail: if the network does not settle because some units have oscillating activation, and if at the end of the run the units representing hypotheses of the rejected theory have activations greater than 0, or even greater than the activations of what should be the winning units. Both failures tend to be the result of having excitation too high relative to inhibition. So long as the value for the excitation parameter is higher than the absolute value of the inhibition parameter, losing hypotheses tend to be rejected and networks tend to settle in around 100 cycles of updating. For all runs reported in this book, the parameter values used were .05 for decay and data excitation, .04 for excitation, and –.06 for inhibition, but many other combinations of values could have been used as long as excitation was less intense than inhibition. The original ECHO (Thagard 1989) tended to need stronger inhibition, since without the principle of competition there were fewer inhibitory links.

CHAPTER 5

Theory Dynamics, Rationality, and Explanation

THE chemical revolution has provided a rich example of conceptual change, and the transition from the phlogiston theory to the oxygen theory can be accounted for by the theory of explanatory coherence. But how typical is the chemical revolution of other major conceptual developments in science? In general, what is the relation between a new theory and the old theory that it replaces? At the theoretical level, Lavoisier's views retained little of the previously dominant phlogiston theory, but we should not immediately assume that this is true of all scientific revolutions. Generalizations about the relations between successive theories must await a broader range of historical cases. The first section of this chapter lays out the possibilities, outlining the different ways in which successive theories can be related to each other. Later chapters will show that scientific revolutions differ in how cumulative they are, although all are cumulative to some extent.

From a philosophical perspective, the topic of the relation between successive theories is crucially relevant to the question of whether scientific revolutions are rational: are new theories adopted because they are objectively superior to old ones, or are they adopted arbitrarily? Section 5.2 applies the perspective of conceptual change and explanatory coherence to three issues relating to rationality. First, theory change in scientific revolutions is often compared to religious conversion. A psychological account of conversion, however, shows that it has little in common with scientific change. Second, sociologists of science have claimed that they can fully explain the development of science without addressing questions of rationality or cognition, but the limitations of their models are easy to see. Finally, Kuhn's recent writings have suggested that the relation between successive conceptual systems is akin to that between languages, but we shall see that the analogy is limited and does not undercut the possibility of rational conceptual change.

The final question addressed in this chapter is a background issue for the theory of explanatory coherence: what is an explanation? I do not have a unified theory of explanation to offer; after outlining the different views of explanation that have been offered by researchers in philosophy of science and artificial intelligence, I describe what a cognitive theory of explanation might look like. Several different approaches to the nature of explanation are compatible with objective establishment of the explanatory relations that TEC requires.

5.1 DYNAMIC RELATIONS OF THEORIES

5.1.1 Theory Replacement

Putting the theory of explanatory coherence from Chapter 4 together with the theory of conceptual change, we get the following picture of the transition from one theory to another:

1. A scientist with a theory embedded in a conceptual system becomes aware of a new theory that competes with the one already held.
2. Although initially skeptical, the scientist sets out to learn more about the new theory, and gradually accumulates its conceptual system and an understanding of its explanatory claims.
3. The scientist comes to appreciate that the new theory has greater explanatory coherence than the old one.
4. The old theory, and its attendant conceptual system, drop into disuse.

Thus transformations in propositional as well as conceptual structure take place. The phlogiston theory had concepts such as the different kinds of earths arranged in kind-hierarchies and part-hierarchies. The oxygen theory had a different set of concepts and a different hierarchical organization. The two theories can be compared, however, because they also contain propositions that are offered as explanations of some of the same evidence concerning aspects of combustion and calcination. When a phlogiston theorist appreciated the greater explanatory coherence of the oxygen theory, the phlogiston propositional and conceptual system became dormant.

The interrelations between conceptual and propositional structure are difficult to visualize, even in simple cases. Figure 5.1 shows partially how conceptual and propositional structure are intertwined in the phlogiston and oxygen theories, both of which explain E1, that wood burns. In the oxygen theory, the proposition O1, that oxygen combines with wood, is part of the explanation. In the phlogiston theory, the proposition P1, that wood contains phlogiston, is part of the explanation. The concepts of wood and phlogiston are linked in the proposition P1, but concepts are also organized in kind-hierarchies. Because of the explanations offered, O1 and P1 each cohere with E1, and O1 and P1 incohere by virtue of the principle of competition. When O1 turns out to be part of the account having the greatest explanatory coherence, P1 is rejected and is no longer used in explanations. Then the conceptual structure at the right side of Figure 5.1 ceases to play any cognitive role.

The picture just presented does not fully answer the question: When one

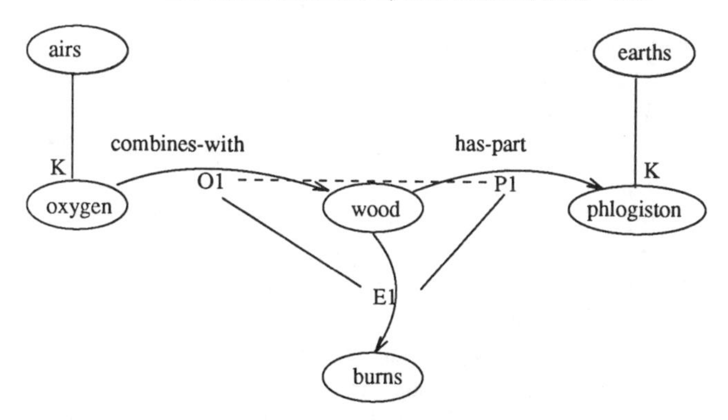

Figure 5.1. Conceptual and propositional structure. O1 is the proposition that oxygen combines with wood. E1 is the proposition that wood burns. P1 is the proposition that wood contains phlogiston. O1 coheres with E1; P1 coheres with E1; O1 and P1 incohere. Lines marked with K indicate kind-relations.

theory replaces another, as the oxygen theory did the phlogiston theory, what is the relation between the old theory and the new one? To what extent is the development of science *cumulative*, incorporating what has gone before? At the most general level, we can distinguish four kinds of relations between successive theories, ordered by decreasing amounts of cumulativeness: incorporation, sublation, supplantation, and disregard.

> If a new theory T2 completely absorbs the previous theory T1, then T2 *incorporates* T1.

This is total cumulation; T2 is just an extension of T1.

> If T2 partially incorporates T1 while rejecting aspects of T1, then T2 *sublates* T1.

The unusual term "sublate" is an English translation of the German verb *aufheben*, which means both to clear away or annul and to preserve. This concept is crucial in Hegel's description of the development of concepts as involving both rejection and preservation. There are many respects in which Hegel's model does *not* fit the development of scientific knowledge (Thagard 1982), but the Hegelian concept of sublation is useful for describing cases where theory choice involves both incorporation and rejection.

> If T2 involves the near-total rejection of T1, then T2 *supplants* T1.

There is then little cumulation in the supersession of T1 by T2.

> Finally, if the adoption of T2 comes about simply by ignoring T1, then T2 *disregards* T1.

By disregard I mean that there is virtually no relation between the two theories and that the adoption of one is not the result of the relative merits of the theories. This relation is rare in the natural sciences, but sometimes occurs in the social sciences when new theories come on the scene like new colors in fashion design.

The incorporation view of science is found among those scientists who suppose that new theories must be consistent with established science. At the other extreme, we find the Kuhnian view that new theories involve the supplanting of previous ones; in Kuhn's most relativist writings, the incommensurability of theories implies that one theory simply disregards the other. Without question, the day-to-day development of physics, chemistry, biology, and some social sciences involves much accretion of new views to old, but scientific revolutions are where one would expect incorporation to fail. But just how much is rejected: do the new theories sublate, supplant, or disregard the old ones?

The questions need to be broadened to distinguish different kinds of cumulation. So far, I have been talking about theory cumulation, but this can be subdivided into *concept* cumulation and *proposition* cumulation. These two are intimately related, since abandonment of concepts goes hand in hand with rejection of propositions employing those concepts, but it is certainly possible to have concept cumulation even though various propositions are rejected. Moreover, even if one theory supplants another, it is possible that it incorporates the evidence for the old theory. *Evidence* cumulation takes place if the evidence that supported T1 also supports T2. In terms of the theory of explanatory coherence, evidence cumulation consists of T2 explaining everything that T1 does. Finally, we can talk of *method* cumulation. Theories often have associated with them methods for designing and performing experiments and for evaluating experimental results. Replacement of theories may sometimes involve replacement of methods as well. Most radically, it is possible that standards of theory evaluation may change with theories. The theory of explanatory coherence would be seriously undermined if historical cases exhibited variability in the principles of explanatory coherence or serious disagreements about what constitutes an explanation. We would then have no method cumulation, and few grounds for saying that the new theory is better than the disregarded old one.

5.1.2 The Relation between the Oxygen and Phlogiston Theories

Now it is crucial to get back to the concrete historical case of the chemical revolution. Discussion in previous chapters suggests that the oxygen theory is best described as *supplanting* the phlogiston theory. The oxygen theory involved the abandonment of so many of the concepts and propositions of the phlogiston theory that it obviously did not incorporate the phlogiston theory. Moreover, Lavoisier's explicit rejection of the existence of phlogiston and the theoretical claims that went with it shows that little of phlogiston theory was preserved in the oxygen theory, so sublation is not the appropriate relation either.

Although there was little theory cumulation in this historical case, it would be a manifest mistake to say that the new theory disregarded the old, since Lavoisier and his colleagues mounted systematic arguments against the phlogiston theory. New experimental evidence arose to support the oxygen theory, but Lavoisier still needed to explain the basic facts of combustion, calcination, and respiration that had made the phlogiston theory so plausible. There thus was considerable cumulation of *evidence* from the phlogiston to the oxygen theories. The ECHO analysis in section 4.3 shows, however, that there was not complete cumulation, since to my knowledge Lavoisier did not attempt to explain the transference of combustibility that the phlogiston theory could explain. But by and large, Lavoisier could explain what the phlogiston theorists did, plus a lot more.

Method cumulativeness is harder to assess. I have not found in the writings of phlogiston theorists such as Kirwan and Priestley any obvious methodological differences with the Lavoisier school. Both camps had similar experimental methods, although Lavoisier developed more careful means of measuring changes in weight. Kirwan and Priestley gave alternative explanations of Lavoisier's experiments, but the alternatives can easily be understood as coming more from doctrinal disagreements than from methodological differences. I see no reason, therefore, to see a substantial methodological rupture in the chemical revolution that would undercut the applicability of the theory of explanatory coherence. Like the oxygen theorists, the phlogiston theorists wanted to explain the evidence, and most of them realized eventually that they would have to abandon their theory.

In sum, in the chemical revolution there was little cumulation of theory, but much cumulation of evidence and method. We shall see in later chapters that this pattern of relations is not found in all other scientific revolutions, which have varying degrees of cumulativeness. To presage later findings: the Darwinian revolution was even less cumulative than Lavoisier's, while revolutions in physics and geology were substantially more cumulative. The oxygen theory represented progress over the phlogiston theory because it explained more and led to experiments that greatly added to the stock of evidence to be explained. The revolution was largely driven by these experiments: the oxygen theory was needed to explain experimental facts such as the weight gain of burning objects that were difficult to fit within the phlogiston framework. Other revolutions, particularly in physics, have been more coherence-driven.

5.2 IMPEDIMENTS TO RATIONALITY

After a talk I once gave about conceptual revolutions, a questioner at the back of the room remarked: "Why, it almost seems that you're suggesting that science is rational!" In an era influenced by Kuhn, Feyerabend, and Derrida, rationality has gone out of fashion. According to Feyerabend (1975), the only legitimate methodological principle is "anything goes." To those philosophers who deny that there is a general scientific method, my theory of explanatory coherence as the driving force of theory replacement will seem anachronistic. But a theory of theory evaluation that can be applied to all the major scientific revolutions is an excellent antidote to irrationalism.

When Kuhn's *Structure of Scientific Revolutions* first appeared in 1962, it lit a bonfire under the philosophy of science and has easily been the most widely read work in the field over the past decades. Credit is due Kuhn (along with others such as N. R. Hanson and Stephen Toulmin) for redirecting philosophy of science away from arid logical analyses of scientific concepts toward greater consideration of the dynamics of science as seen in its history. Kuhn helped to revitalize the study of conceptual change in science that had lain dormant in Anglo-American philosophy since the great nineteenth-century thinkers William Whewell and C. S. Peirce. But the aspect of Kuhn's monograph most distressing to philosophy was its implication that science was something less than rational. When one "paradigm" superseded another, it brought such a new way of looking at the world that Kuhn even asserted that "the world changed." Paradigms, said Kuhn, are *incommensurable*, which many commentators took to mean that scientific theories cannot be rationally evaluated to see which is better. In the postscript to the second edition of his book in 1970, Kuhn modified his position considerably, and subsequent writings have backed off more and more from the suggestion that scientific revolutions are irrational occurrences (Kuhn 1977, 1983).

The purpose of this section is to defend the rationality of scientific revolutions by showing how various impediments to the application of the standards established in my theory of explanatory coherence can be overcome. The first subsection considers the analogy, popular among those who are concerned to downplay the rationality of conceptual revolutions, between acceptance of new conceptual systems and religious conversion. Exhibiting the weakness of the analogy provides a contrasting lesson about the comparative rationality of science. I characterize religious conversion as a kind of motivated inference, in which belief change is governed by personal goals. Scientists, like people in general, are not immune to motivated inference, but social structures keep it under control. The second subsection challenges the claims of some sociologists concerning the social determination of scientific knowledge.

Kuhn's (1983) more recent discussions of incommensurability of theories emphasize similarities between scientists facing a new theory and language learners facing a strange new language that they want to be able to translate into their own. Section 5.2.3 explores the analogy between language learning and acquisition of conceptual systems. We shall see that scientific conceptual systems are in some respects easier and in other respects harder to acquire than new languages. The key point to be made is that even though the need to interpret and acquire a new conceptual system is an impediment to rational theory evaluation, it is one that can be overcome.

5.2.1 Conversion and Motivated Inference

Kuhn (1970) compared the shift from one paradigm to another to religious conversion, and many other authors have used conversion as a metaphor for what occurs in science (Cohen 1985, ch. 30). In some respects, the conversion metaphor is useful, since it suggests that adoption of a new theory can require a major transformation. But the metaphor is dangerously inaccurate: it is inaccurate because the cognitive processes underlying religious conversion are very different from those underlying adoption of a scientific theory; and it is dangerous because it suggests that the transformation is irrational.

William James (1982) provided the classic description of conversion more than a hundred years ago. He emphasized the emotional aspect of religious conversion. For example, he described the conversion of young people as following symptoms of a sense of incompleteness and imperfection, brooding, depression, morbid introspection, and sense of sin; the conversion replaces these with a happy relief and confidence in self. Converts, from the ordinary to the famous such as St. Augustine and St. Paul, see themselves as struggling away from sin. Converts often see themselves as passive spectators to an astounding process performed on them from above. James described the feelings that typically attend the conversion experience: a sense of higher control, the loss of all worry, a beautification of the world, and ecstasy. More recent investigations such as Salzman's (1953) have also remarked on how sudden conversions are often preceded by severe emotional strains.

If these accounts are correct, then religious conversion should be thought of as the acquisition of a belief system driven by intense emotional involvement. Conversion takes place as a solution to the emotional problems of the person converting, and has little to do with realizations of the plausibility of the doctrine that may be accepted. Conversion is thus better described as emotional reorientation toward a perceived "higher power" than as belief revision or conceptual change, although revision and change can result from the reorientation.

There is no evidence that conceptual change in scientists is either driven or accompanied by the emotional intensity found in religious conversion. Scientists may be distressed by a new theory, especially if it suggests that their old way of thinking about things is endangered and obsolescence looms, but they do not adopt it as a way of alleviating that distress. It is unfortunate that we do not have more historical data on the adoption of new theories by scientists. From their writings, one can infer roughly when they adopted a new theory, but I know of no diaries or other contemporaneous reports that mark the adoption of new theories. What descriptions there are, however, of scientists' transitions do not record the experiences of ecstasy that converts reportedly undergo. Undoubtedly scientists do experience excitement, particularly if the new conceptual framework suggests interesting new work to do, but this is mild compared to the convert's emotional transformation. In short, religious conversions are largely emotion-driven, unlike adoption of theories by scientists.

This is not to say that scientists are unemotional and unmotivated. Most successful scientists work with great emotional intensity on projects that they believe in, and only a fool would say that scientists' motivations do not include such nonepistemic goals as fame and professional advancement. The question is: how much do these goals bias their work? Kunda (1987, 1990) has conducted a series of experiments on cases where people's inferences are biased by their personal goals. One of her studies shows that people tend to generate theories in a self-serving manner that supports the optimistic view that their own attributes predict good life outcomes but not bad ones such as divorce. Subjects whose own mothers had been employed outside the home viewed having had an employed mother as better for marriage and having had a nonemployed mother as worse for marriage than did subjects whose mothers had not worked outside the home. In another study, students who were planning to go to graduate school were more likely than others to see their own attributes as predictive of success in graduate school. In a third study on evidence evaluation, subjects who were heavy coffee drinkers were found to be less willing to believe an article about the negative health effects of caffeine.

How motivated are the inferences made by scientists in changing their beliefs? Kunda's work does not support the claim that *all* inference is motivated inference. When motivational pressures are reduced in the caffeine experiment by making the health consequences of caffeine seem less severe, heavy consumers of coffee were no less convinced by the evidence than were light consumers. Kunda concludes that "people do engage in motivationally directed inferential processes, but only when the levels of motivation are high. Even when motivation is high, they do not completely ignore negative evidence; they are responsive to it, but not as responsive as they might have been in the absence of a motivational involvement" (Kunda 1987, 644f.).

It is easy to expand ECHO into Motiv-ECHO to take into account the personal motives of a reasoner. We can add the following two principles of motivated explanatory coherence:

Principle M1: Motivational coherence.
If P contributes to a person's goal G, then P and G cohere.
Principle M2: Goal priority.
A goal G has a degree of acceptability on its own.

Translated into ECHO in a way analogous to Principle 4, Data Priority, Principle M2 provides activation to units representing goals. In accord with Principle M1, these goal units then spread activation to units representing those propositions that would contribute to the goals. Motiv-ECHO can thus be used to model cases of theory evaluation biased by personal goals.

Scientists' motivations fall into several categories. First, scientists have personal goals such as fame, professional success, and financial gain. Second, they may have group-related goals, such as wanting their research teams to prevail over competing views. Third, scientists may have national or political goals, preferring ideas associated with their countries, ethnic groups, or social classes to be dominant. On some sociologists' interpretations discussed in the next section, these kinds of goals are the primary determinants of the views adopted by scientists. However, in contrast to ECHO's historical successes, there is little evidence to support a Motiv-ECHO model of theory choice. The next section makes this point by discussing the antirationalist views of some recent sociologists of science. It turns out, surprisingly perhaps, that scientists can be quite rational, preferring theories with greater explanatory coherence. The only serious impediment to recognizing that someone else's theory has greater coherence is the problem of sufficiently acquiring the other theory's conceptual system to make possible appreciation of its explanatory claims. Motiv-ECHO shows that the cognitive/computational approach is not restricted only to models of human rationality: we can also work out in detail what people appear to be doing when they make irrational judgments. The question of the rationality of science then comes down to: Does the cognitive model that gives the best account of the widest range of scientific thinking embody rationality? Some sociologists have given this question a negative answer.

5.2.2 *Sociological Explanations*

Since the emergence of the "strong programme" in the sociology of science more than a decade ago, sociologists and philosophers of science have been at loggerheads (Brown 1981). Whereas philosophers have traditionally attempted to explain developments in science using models of rationality, advocates of the strong programme claim that explanations of scientific change can be purely sociological, emphasizing the "interests" of scientists rather than evidential grounds for acquiring beliefs. The positions of philosophers and sociologists are well illustrated in principles that have been proposed by Larry Laudan and David Bloor. Laudan considers, but does not endorse, restricting the sociology of science by an *arationality assumption*: "The sociology of knowledge may step in to explain beliefs if and only if those beliefs cannot be explained in terms of their rational merits" (Laudan 1977, 202). Contrast this assumption with Bloor's principles of *impartiality* and *symmetry*, according to which a proper account of the sociology of science will use the same types of causes to explain beliefs regardless of whether they are true or false, rational or irrational, successful or unsuccessful (Bloor 1981). Whereas Laudan's principle says that sociology is relevant only to the residue of scientific practice that remains after models of rationality have been applied, Bloor sees no role at all for such models and clears the way for sociological explanations. Both of these views strike me as insupportably dogmatic. Against the arationality assumption, I want to maintain that there is no advance reason why we should give special preference to explanations of belief change in terms of rationality. Instead, we should attempt to give the best explanation we can of particular scientific episodes, judging for each episode whether social or rational factors were paramount. Against Bloor, however, I argue that in many cases the best explanation comes from a model of rational belief change such as ECHO.

What is the best explanation of changes such as the adoption of the oxygen theory? There is no evidence of alteration in the personal, group, or national motivations of the scientists in these periods. If theory acceptance were driven by such motivations, then change in the theory that is accepted would have to depend on some change in motivation that has not been recorded. On the other hand, there is ample evidence in each of these revolutions of an abundance of new experimental results and arguments concerning their explanation. Lavoisier did extensive experiments to show, for example, that burning substances gain weight, a fact hard to explain on the view that in combustion phlogiston is given off. Later chapters will describe more cases that have proven highly amenable to analysis using the theory of explanatory coherence.

Undoubtedly, proponents of the strong programme can find motivational elements in these cases. When we explain, for example, the acceptance of quasi-Darwinian views about survival of the fittest among nineteenth-century American capitalists such as Andrew Carnegie, motivational, ideological explanations should doubtlessly prevail. But the historical record concerning Lavoisier, Darwin, and plate tectonics provides no reason for considering motivational factors to be the main causes of theory change. Rather, in each of these revolutions, the historical evidence suggests that a rational model such as one provided by my theory of explanatory coherence is the most appropriate.

Part of this preponderance is due, curiously enough, to social factors. Even if a scientist is driven by personal motivations of success and fame, he or she has to present research to the rest of the scientific community in terms of its experimental and theoretical merits. Through the process of peer review, personal motivations tend to be canceled out. In science on the international scale at which it is now practiced, group and national motivations also tend to cancel out. Because of an institutional commitment of science to experimental evidence and explanatory argument, science as a whole is able to transcend the personal goals of its fully human practitioners who acquire the motivation to do good experiments and defend them by rational argument. Although cases of fraud occasionally arise, the social institutions of science do a fair job of maintaining scientific rationality.

In sum, neither religious conversion nor motivated inference provides as good a cognitive model of scientific theory acceptance as does my theory of explanatory coherence, so on empirical grounds we can judge conceptual revolutions to be rational. My argument should not be construed as denying the relevance of sociology to the understanding of scientific development. There are fascinating questions about scientific organizations and national styles that a full account of science must address. McCann (1978) points out numerous interesting social aspects of the chemical revolution. Dissemination of the oxygen theory began with Lavoisier's immediate associates, only slowly spreading out beyond his circle to the rest of France, and slower still to other countries. Part of the problem in England was that chemistry was decentralized, with little activity at Oxford and Cambridge. In contrast, the leading French chemists were concentrated in Paris, facilitating the personal contacts that undoubtedly aided transmission of the oxygen theory. My cognitive account of conceptual change and theory acceptance is obviously not intended to be the whole story of scientific development, but it shows that a purely sociological story would never do either. I am certainly not claiming that the sociology of science *reduces* to the psychology of science; explanation can fruitfully proceed at both levels. (For further discussion of the roles of cognitive and sociological explanations, see Slezak 1989.)

5.2.3 Translation

Kuhn's doctrine of the incommensurability of theories has caused no end of philosophical consternation. The standard refutation of it is that if two theories were not commensurable—if they really were about different worlds—then they were not really competing, so no problem of rationality arises. Kuhn has clarified his position to state that although theories may be incommensurable, that does not make them incomparable, even though they lack a common language. "The claim that two theories are incommensurable is then the claim that there is no language, neutral or otherwise, into which both theories, conceived as sets of sentences, can be translated without residue or loss" (Kuhn 1983, 670). According to Kuhn, acquiring a new theory is like acquiring a second language, which does not guarantee that you will be able to translate between the languages.

The analogy between theory acquisition and language acquisition fits well with the account given in Chapter 3 concerning the acquisition of conceptual systems, but the translation question is potentially damaging to the theory of explanatory coherence. If there is no translation possible between the oxygen theory and the phlogiston theory, how can one assess their comparative explanatory coherence? The answer to this question will require a closer look at the nature of translation and at the amount of translation required to generate judgments of explanatory coherence.

To begin with, we must reject Quine's (1960) philosophically influential view of the indeterminacy of translation. According to Quine, there are many different nonequivalent ways to translate from one language to another and no fact of the matter as to which of these ways is correct. His basic notion is sameness of stimulus meaning, which is possessed by two speakers if the same sensory stimulations prompt assent in both. Translation manuals should ideally identify terms with the same stimulus meaning, and translation becomes indeterminate to the extent that translation manuals are not based on direct links with nonverbal stimulation. Quine's views on translation are much like the discredited behaviorist theories of B. F. Skinner, which will be discussed in Chapter 9. Quine and Skinner err in taking conditioning to nonverbal stimuli as the core case of language learning. From a cognitive perspective, language learning consists of acquiring complex mental representations of the grammar and lexicon of a new language. Translation then involves the ability to use these representations to transform terms and sentences of one language into terms and sentences of the other. The meaning of a term may be related to nonverbal stimuli, but it is hardly determined by them. Following Harman (1987), I prefer the theory of *conceptual role semantics*, according to which the meanings of symbols are determined by their functional role in thinking, including perception and reasoning. From this perspective, it is folly to talk of the meaning or content of a term independent of a general theory of cognitive processing that determines functional roles. Rather than stimulus meaning, one should be looking at the cognitive operations that employ a representation of the term.

Let us now systematically compare, from a cognitive perspective, acquiring a new system of scientific concepts with learning a new language such as French or Swahili. To speak a language, one must acquire its grammar and its vocabulary, or lexicon. Fortunately, acquiring a scientific conceptual system does not require learning a largely new set of syntactical relations. No one would claim there are differences in syntax between the oxygen and phlogiston theories. In contrast, the lexical differences are substantial: each theory has terms such as "phlogiston" and "oxygen" that do not appear in the other, and we saw in the last chapter that the lexicons of the two conceptual systems are organized very differently. Thus as far as the scientific lexicon is concerned, translation is indeed problematic.

But acquiring the lexicon of a new natural language is no day at the beach either. Larson (1984) provides numerous fascinating examples of mismatched lexical systems. For example, the Tzeltal language of Mexico has no word corresponding to the English *carry*, but has more than a dozen words describing particular kinds of carrying. Many languages lack English color words, so that in Mbembe one word corresponds to English words for red, orange, and yellow. Words have multiple senses, and there is often no neat correspondence between the multiples in one language and those in another. The English word "key" can correspond to Spanish words "llave," "clave," or "tecla," while "llave" can in turn correspond to English words "key," "faucet," or "wrench." The Munduruku language has a word for *high-living* things that embraces birds and monkeys, but does not have separate words for those concepts. Different terms can have different emotional connotations, which causes great problems in translating works of literature (Chukovsky 1984).

Kind-hierarchies and part-hierarchies differ between languages. Some South American languages have ten different words corresponding to the English "banana," indicating more branches in the kind-hierarchy. Slavic languages do not have separate words for *arm* and *hand*, indicating fewer part-whole discriminations than are made in English. Differences such as these, where one language makes more discriminations than the other, are not the most severe mismatches. The English division of *sibling* into *brother* and *sister* differs markedly from the tripartite division of Javanese into *older brother*, *older sister*, and *younger sibling*. In Hungarian, a chicken is not a kind of bird, but instead is a kind of animal (Heltai 1988), and the same is true in modern Hebrew: Israelis look confused if you ask them if a chicken is a bird, a question that generates an immediate positive response in a speaker of English, French, or German. According to Heltai, the differences in the kind-organizations of English and Hungarian account for a substantial number of errors found in Hungarians learning English.

Grammatically, acquiring a scientific conceptual system is unproblematic; perhaps there are exotic languages with radically different structures that imply different ontologies, but modern science has had no difficulty working primarily in the similar languages Latin, English, French, and German. Lexically, however, revolutionary conceptual change is both easier and harder than vocabulary acquisition in natural language. In the first place, there is obviously enormous lexical overlap even between conceptual frameworks as different as the oxygen and phlogiston theories. By the time of Lavoisier's textbook of 1789, this was less true, but such changes came about gradually. In his early writings, Lavoisier used many concepts that were completely familiar to phlogiston theorists: *sulfur*, *phosphorous*, *gold*, etc. The experimental apparatus used by oxygen theorists and phlogiston theoriests were similar, and they often repeated each other's experiments. There was thus far more lexical overlap between the languages of oxygen and phlogiston theories than between the languages of English speakers and South American Indians.

But natural language translation does not seem to encounter as severe problems with ontology, branch switching, and hierarchy redefinition as occur in conceptual revolutions. The difference between oxygen and phlogiston theories is not just that one has the term "oxygen" and the other has the term "phlogiston"; the theorists disagree about whether phlogiston exists. Language learning may perhaps involve some changes in placements in kind-hierarchies, as when a Hungarian acquires the belief that a chicken is a kind of bird. But this does not involve the rejection of the belief that a chicken is a kind of animal, since birds are still animals; it is not branch-jumping. Whorf (1956) claimed that the Hopi language carried with it a different metaphysics of space and time, which suggests that redefinition of the part-hierarchy may be necessary in learning a Hopi language. Whorf's hypothesis is not, however, considered well-established by psycholinguists.

The upshot of this discussion is that acquiring a scientific conceptual system is like learning a second language in some respects, but that language learning requires less radical changes. Hundreds of millions of people are bilingual, and many of them can move easily between the two or more languages that they know (Grosjean 1982). Translation between natural languages is difficult, but people do it all the time. The same is true of evaluating theories embedded in different conceptual frameworks: it is difficult, but people manage to do it.

Most important, complete translation is *not* required for objective judgments of explanatory coherence. I agree with Kuhn that no general, tight, perfect translation "without residue" is possible between conceptual frameworks as different from one another as the oxygen and phlogiston theories. The two conceptual systems do not map onto each other at all neatly, as shown most acutely by the differences in the kind-hierarchies and part-hierarchies. But *complete translation is not necessary for judgments of explanatory coherence*. Recall what kinds of statements are necessary for input to the program ECHO. To compare an opposing theory with one's own in a reasonably objective way, one needs to have sufficient grasp of another theory to appreciate its claims about what explains what and what the data are. Considerable work may be necessary to acquire the lexicon and conceptual system of the opposing theory, but one need not be able to translate between them. One needs to be able to make judgments such as: "According to the phlogiston theory, combustion is explained by phlogiston being given off. This competes with my claim that combustion is explained by absorption of oxygen." It is not necessary to identify particular statements of the opposing theories as contradicting each other, since competition in ECHO can arise merely from two hypotheses explaining the same evidence. For the developer of the new theory who grew up with the old theory, noting the explanations of the old theory should be no problem. Lavoisier had the oxygen theory because he created it, but he still understood the language of the phlogiston theory.

The problem lies on the other side. Can the proponents of the old theory learn enough of the language to be able to appreciate the explanatory coherence of the new theory? The historical record indicates that they can. We saw, for example, that Kirwan went over to the oxygen side. Given the conceptual differences between the two theories, it is not surprising that it sometimes took years to appreciate the purported explanations of the other view. Contradictions do not seem so difficult to appreciate: a theorist can usually recognize that the proponent of another theory is offering an alternative. In most of the revolutions examined in this book, there was large agreement about what could qualify as an explanation. The major exception is probably the debate between Newton and the followers of Descartes, who rejected Newtonian gravity as an occult force with no explanatory power.

The key question is whether the proponents of the two theories could agree on what counts as data. The doctrine that observation is theory-laden might be taken to count against such agreement, but that doctrine only undermines positivist views that there is a neutral observation language sharable by competing theories. My position requires only that proponents of different theories be able to appreciate each other's experiments. In the short run, there are undoubtedly impediments to this, as scientists are more likely to find fault with experiments that do not fit with their own views. All this, however, can come out in the wash as experiments are replicated and arguments reconsidered. Experiments are complicated, and their conduct is hardly independent of theoretical considerations. But the results of experiments are not determined by theory: the world does intervene. In eighteenth-century chemistry, much repetition and replication of experiments took place, with relatively little disagreement about particular experimental results.

In sum, acquiring a new conceptual system is somewhat like acquiring a new language, but people do it all the time, and language acquisition and translation are only temporary impediments to the rational application of explanatory-coherence considerations to theory evaluation.

5.3 PHILOSOPHICAL AND COMPUTATIONAL MODELS OF EXPLANATION

The theory of explanatory coherence in Chapter 4 depends on the existence of an objective notion of explanation. ECHO simulations require input of the form *H explains E*, presupposing that a computational system could be developed that could generate such expressions. What we need is a theory of the nature of explanation and a computational implementation that could construct and identify explanations. Such a theory and implementation are not yet available, but a survey of philosophical and computational research on explanation can help point to how they might arise.

For several decades, explanation has been a central topic in the philosophy of science. Much more recently, the concept of explanation has become important in artificial intelligence. A comparison of the two lines of research makes clear the close conceptual connections between philosophy and artificial intelligence, and highlights the diversity of approaches to explanation. I shall argue that the different approaches reflect the inherent diversity of explanations.

There are at least six different approaches to the topic of explanation that have found favor in philosophy, and at least four of them have corresponding ideas in AI. My claim is that these approaches are best conceived not as alternative theories of explanation, but as views of different aspects of explanation. Investigations of explanation, both philosophical and computational, have resembled the proverbial examination of an elephant by a group of blind Indians, each comprehending only part of the whole. I shall argue that what is needed is a theory of explanation that integrates all the aspects of explanation—deductive, statistical, schematic, analogical, causal, and linguistic. Such a theory can be given in the context of what in AI is called a *cognitive architecture*, a general specification of the fundamental operations of thinking. Researchers in AI and cognitive psychology have been attempting to build cognitive architectures, usually embodying both a theoretical specification of basic representations and processes and a concrete computational model. The development of a theory of explanation can be part of the development of such an architecture.

5.3.1 Strands of Explanation

The views of the nature of explanation that have proliferated in philosophy can be divided roughly into six camps, and AI research falls into approximately the same categories. Explanation has been viewed by different researchers as deductive, statistical, schematic, analogical, causal, and linguis-

Table 5.1
Approaches to Explanation

Strand	*Philosophy*	*Artificial Intelligence*
Deductive	Hempel (1965)	Mitchell et al. (1986) O'Rorke et al. (1990)
Statistical	Salmon (1973)	Pearl (1988)
Schematic	Kitcher (1981)	Schank & Abelson (1977) De Jong et al. (1986)
Analogical	Campbell (1957) Hesse (1966)	Schank (1986) Thagard et al. (1989)
Causal	Salmon (1984) Miller (1987)	Pearl (1988) Peng & Reggia (1990)
Linguistic/ pragmatic	Achinstein (1983) van Fraassen (1980)	

tic. This section quickly reviews the philosophical and computational research within the various perspectives, and argues that the approaches are fundamentally complementary rather than contradictory. I shall argue that explanation is a complex process that can include subprocesses corresponding to the six views of explanation. We shall see in later chapters that explanations by major scientific theories incorporate various deductive, statistical, schematic, analogical, causal, and linguistic aspects as interwoven strands. As a preview to the discussion, Table 5.1 provides a summary classification of the research to date on explanation.

EXPLANATION AS DEDUCTIVE

By far the most influential view of explanation in philosophy has been the *deductive-nomological* model of Carl Hempel (1965; for a recent survey of philosophical research on explanation, see Kitcher and Salmon 1989). On Hempel's view, an explanation is a deductive argument in which the premises include general laws and the *explanandum*—what is explained—is a statement that follows from the premises. Examples of deductive explanations are commonly found in physics. For example, when astronomers explain the occurrence of an eclipse of the sun, they do so by deriving the occurrence from general laws of motion and specific information about the location of the sun, earth, and moon. Deductive explanations yield predictions: if you can explain something, then you could have predicted it.

Philosophers have argued that the deductive-nomological model is neither necessary nor sufficient for explanation. It is not necessary, because fields such as evolutionary biology contain explanations in which laws are vague and deductions are lacking. We can give evolutionary explanations of how humans came to have large brains, but we do not know nearly enough to do so deductively or to predict the future course of evolution. And deduction from laws is not sufficient for explanation either, as the following example shows (Bromberger 1966). From the law of the rectilinear propagation of light, basic trigonometry, and the length of a flagpole's shadow, we can deduce the height of the flagpole; but we have hardly explained why the flagpole has the height that it does. Under normal circumstances, the height of the flagpole explains the length of the shadow, but not vice versa.

Such examples are devastating to the deductive-nomological model as an *analysis* of the concept of explanation, but they do not undermine the fact that the model captures important aspects of a large class of explanations. Explanations by Copernicus, Newton, Einstein, and other physicists typically have the character of mathematical derivations involving general laws. Many of these cases fit Hempel's model well, since they involve deduction of what is explained from general laws. Other cases of explanation, however, seem to have a different character.

Deductive views of explanation have also been prominent in AI. Much recent work in the field of machine learning has concerned *explanation-based* learning. Most early work in machine learning concerned unguided learning from examples; in philosophical terms, such learning corresponds to inductive generalization. "Explanation-based" learning takes place when a system has rich enough knowledge of a domain that it can learn even from a single example by explaining to itself why the example is an example of the concept under study. DeJong and Mooney (1986) describe how a system might learn from a single example that cups are light and have handles. The training example is C, a cup that is red, light, and has a handle. Background knowledge includes the generalizations: if something is an open vessel, is stable, and liftable, then it is a cup; if something is light and has a handle, then it is liftable. The task is to produce a generalization about cups from the training example C and background knowledge, without having to collect myriad examples of cups. As part of the explanation of why C is a cup, the system uses the information that C is light and has a handle to infer that it is liftable, and uses the fact that it is liftable to infer that it is a cup. Since being light and having a handle are integral to the explanation of why C is a cup, the system generalizes that cups are light and have handles.

I shall not go into the details of the additional constraints required to make this kind of inference reasonable; my concern is not with explanation-based learning as such, but only with the view of explanation that is implicit in such models. Here we find an interesting divergence among AI researchers. For Mitchell, Keller, and Kedar-Cabelli (1986), explanation is clearly deductive: an explanation that C is a cup is a *proof* that it is. Much AI research has been done on deductive systems, usually called *theorem provers*. If deduction is viewed as the central operation of a system, then explanation will also be an essentially deductive notion. I shall shortly describe the alternative, schematic, view of explanation and explanation-based learning advocated by DeJong and Mooney (1986).

Theorem provers are not the only deductive systems in AI. Production systems, pioneered by Newell and Simon (1972), are widely used, both as models of human cognition and as expert systems. A production is a general IF-THEN rule, in which a successful match of the IF part against a set of accepted propositions leads to inference of the THEN part. For example, a rule that says that if x is a dog, then x has a tail, would match against the proposition that Fido is a dog, producing the conclusion that Fido has a tail. Such inferences might seem to philosophers like a trivial combination of logical rules of universal instantiation and *modus ponens*, but innumerable interesting issues arise when one addresses the procedural question of when to fire rules in a large system (see, for example, Buchanan and Shortliffe 1984). In practical computational systems, the logician's handy assumption of deductive closure is disastrous; effective ways must be found to control what inferences are made. Rule-based systems are usually discussed in the context of problem solving: given a set of starting conditions, rules must be fired to lead to the accomplishment of a set of goals. But there is a natural translation of explanation in the contexts of such systems. If the goals are taken to be a set of facts to be explained, then an explanation consists of using rules to derive these goals using known information and rules that are matched. If the rules are universal generalizations, as they often are, then the explanation is deductive. Elsewhere I have described a computational model in which explanation is taken to be in part a kind of rule-based problem solving (Thagard 1988).

As we saw in Chapter 3, C. S. Peirce used the term "abduction" to refer to the formation of explanatory hypotheses. Implicit in most AI research on abduction is a conception of explanation as essentially deductive; O'Rorke and his colleagues (1990), for example, use deduction in the programming language PROLOG to perform explanations. So in AI as in philosophy, explanation is often viewed as deductive, but this is only part of the story.

EXPLANATION AS STATISTICAL

Wesley Salmon (1970) rejects the assumption of Hempel's model that explanations are arguments aimed at showing why what is explained was to be expected. Instead, he describes explanation of an event as essentially concerned with describing what factors are statistically relevant to the occurrence of the event. We might not be able to predict that an individual smoker will get cancer, but the statistical relevance of smoking to cancer provides us with an explanation. In contrast, Hempel saw statistical explanation as much like deductive-nomological explanation, in that it consists of an argument from statistical generalizations showing that the event was probabilistically to be expected. Railton (1978) showed how some kinds of statistical explanation could be understood in terms of deductive derivation from statistical laws.

Although an exact characterization of the nature of statistical explanation has proved difficult to achieve, there can be no denying the importance of statistical explanation in science and everyday life. What is needed is a theory that will both adequately account for typical statistical explanations and also cohere with other kinds of nonstatistical explanation. On Salmon's current view, statistical relevance is just one aspect of explaining things in terms of the causal patterns of the world.

AI researchers have also been concerned with statistical explanation, although not usually characterized as such. Numerous systems have been developed for medical diagnosis, which can naturally be understood as a process of finding explanations for symptoms associated only statistically with the diseases that cause them. The MYCIN project at Stanford developed a method of dealing with uncertainty that was not probabilistic; it used *certainty factors* that deviate from probabilities in numerous ways, for example in varying between 1 and –1 (Buchanan and Shortliffe 1984). Other models of diagnosis by Peng and Reggia (1990) and Pearl (1988) have used probability theory, but these researchers are fully aware that causal reasoning is not just a matter of calculating probabilities.

It is obvious both in science and in ordinary life that explanation is often statistical. Any theory of explanation will have to incorporate the statistical strand, but it remains unclear how to tie this strand together with the deductive and other aspects of explanation.

EXPLANATION AS SCHEMA APPLICATION

Deductive and statistical aspects of explanation have been susceptible to formal analysis, but there are less formal strands of explanation that cannot be ignored. Some researchers in both philosophy and AI have asserted that to explain a fact is to fit it into a pattern or system of beliefs. In a recent lecture, Smart (1989) describes how the general conception of explanation as fitting into a system subsumes a variety of other conceptions. For example, the deductive-nomological model uses deductive relations to relate an explanandum to the beliefs from which it follows logically. He also discusses how analogies fit beliefs into a system in a different way discussed below. An additional way of fitting, not discussed by Smart, involves applying *schemas*, organized patterns, to a description of what is to be explained. Following the emphasis of Friedman (1974) on the importance of explanations providing unification of disparate phenomena, Kitcher (1981) describes how theoretical explanation consists of providing unifications of phenomena using problem-solving schemas. Kitcher describes patterns of explanation used repeatedly by Darwin and

Newton in applying their theories to a wide range of phenomena. On this view, explanation is not a matter of showing that an event is to be expected for deductive or statistical reasons; rather, explanation involves unifying a range of phenomena using schemas.

In artificial intelligence, the view of explanation as schema application has been championed by Schank and Abelson (1977). Their *scripts* are complex schematic representations that describe typical events. Explaining events is a matter of applying an appropriate script to them. For example, a script for *restaurant* contains a description of the typical sequence of events in a restaurant: being seated, ordering, eating, paying, and so on. So an explanation of why someone left money on the table before leaving would come by matching the restaurant script to the situation and instantiating the tipping part of the script by the episode of leaving money. The explanation is not deductive or statistical, but comes by fitting the event to be explained using a schema for typical restaurant behavior. Abelson and Lalljee (1988) discuss explanations as applications of *explanation prototypes*, which are generalized versions of the explanation patterns of Schank (1986).

Churchland (1989) has also espoused a view of explanation as prototype application, but he understands prototypes as patterns of activation distributed over units in connectionist systems. He claims that explanatory understanding is analogous to perception and consists in the activation of a specific prototype vector in a well-trained network. This view would be more appealing if connectionist models existed that were capable of the kinds of high-level reasoning that normally contributes to explanation. Connectionist systems have been successful in some kinds of perceptual and learning tasks, but have not yet been extended to cases that we normally describe as explanations. The connectionist prototype view is a suggestive way of thinking of schema application, but its contribution to the understanding of explanation is at best metaphorical until someone builds a connectionist explanation program. Perhaps someday we will have connectionist/neurological models that subsume all the strands of explanation I have been discussing, but such models will require substantial additions to the power of current models (see section 9.3.4).

Earlier I described how one view of explanation-based learning in AI assumes a deductive model of explanation. In contrast, DeJong and Mooney (1986) have a schema-application model of explanation akin to Schank and Abelson's. Mooney and DeJong (1985) describe a system that forms a new schema for *kidnap* as the result of an attempt to fit a description of an abduction and subsequent bargaining into a knowledge base of hierarchically organized schemas.

Obviously, there are many details that need to be filled in as part of a model of explanation as schema application. We need to specify what the schemas consist of, how they are organized in the total conceptual system, and how they are applied to provide explanation of particular events. Nevertheless, there is much appeal in the view that explanation is schema application, in both science and in everyday life. But one should not try to defend a view that explanation *is* schema application rather than deduction, but instead try to see how both functions can be performed within the same cognitive architecture.

EXPLANATION AS ANALOGICAL

In ordinary life, people often use analogies to explain events. Ask parents a question about a child malady and their answer will often draw on experience with their own children being sick rather than more general information. Dolnick (1989) used a revealing analogy in a recent article on pandas. After describing their digestive and reproductive inadequacies, he raised the question of how pandas have managed to survive for millions of years. He remarks that in evolution, as in television, it is not necessary to be good; you just have to be better than the competition. In philosophy of science, the use of analogies in explanation has been advocated by Campbell (1957) and Hesse (1966). In his discussion of theories, Campbell rejects the suggestion that analogy between gases and swarms of elastic particles is incidental to the dynamical theory of gases. He asserts that "analogies are not [just] 'aids' to the establishment of theories; they are an utterly essential part of theories, without which theories would be completely valueless and unworthy of the name" (Campbell 1957, 129). In explanation, analogies are essential in relating strange phenomena to more familiar phenomena. Campbell overestimates the extent to which explanation is reduction to the familiar, and exaggerates the generality of the use of analogy in explanation; but he is undoubtedly on target in seeing that sometimes analogies are very important to scientific explanation. Darwin, for example, frequently used the analogy between natural and artificial selection in his explanations (Chapter 6). Thus a full theory of explanation will undoubtedly have to accommodate the analogical component of some explanations.

In AI, Schank (1986) and his colleagues have considered the use of analogies in explanations. When you explain an event, you store in memory the explanation as an *explanation pattern* that can then be used to explain analogous events. Schank describes a program that generated explanations for why the race horse Swale died by analogy to other events, such as the death of runner Jim Fixx. Analogical explanation is viewed as an addition to the schema-based explanations provided by more general structures such as scripts. Analogical explanation can contribute toward schema application, since an analogy between two cases can be used to generate a schema that generalizes them both (Gick and Holyoak 1983).

AI now has a body of research on *case-based reasoning*, although more of it is concerned with analogical problem solving than with analogical explanation (Hammond 1989). Other AI work on analogy has concerned how two analogs can be put in correspondence with each other (Falkenhainer, Forbus, and Gentner 1989; Holyoak and Thagard 1989). In explaining an unfamiliar phenomenon such as the structure of the atom in terms of a familiar one such as the solar system, it is crucial to be able to work out the systematic correspondences between the two analogs. One needs to notice that the location of a proton at the center of the atom corresponds to the place of the sun at the center of the solar system. Analogies are often used in explanations in chemistry and chemical education (Thagard, Cohen, and Holyoak 1989).

The AI view of analogical explanation as a supplement to other explanatory techniques is more reasonable than Campbell's generalization that theoretical explanation is always analogical. But much remains to be done to build a system that integrates analogy with other strands of explanation.

EXPLANATION AS CAUSAL

From the logical-positivist perspective that spawned the deductive-nomological model of explanation, the notion of *cause* was of scant philosophical respectability. In contrast, a growing band of philosophers contends that causality is a central part of explanation. According to Salmon (1984), to explain an event is to show how it fits into the causal structure of the world. For Miller (1987), an explanation is an adequate description of underlying causes. Hausman (1982) and others have argued that the asymmetries of explanation, for example that flagpole heights explain shadow lengths but not vice versa, should be understood in terms of causal relations. While some philosophers such as Achinstein (1983) have denied that all scientific explanation is causal, citing causes is clearly an important part of a great many explanations.

Causality has been more peripheral to recent AI discussions of explanation, probably because it is not at all clear how to get a deep representation of causal knowledge. *Qualitative physics* is an active research topic in AI concerned with computational models of simple physical systems and issues of causality frequently arise (Bobrow 1985). Researchers on analogical mapping agree that causality is very important for putting analogs in correspondence with each other, but provide no theory of causality (Holyoak and Thagard 1989; Falkenhainer, Forbus, and Gentner 1989). Schank and his colleagues view schema application and the use of explanation patterns as causal reasoning, but also have little to say about what makes it causal. If causality is as important to explanation as many philosophers think it is, then AI systems are going to require much more sophisticated notions of causality than they currently possess.

LINGUISTIC AND CONTEXTUAL ASPECTS OF EXPLANATION

For the sake of completeness in my quick summary of the recent philosophical literature, it is necessary to mention the influential contributions of Achinstein (1983) and van Fraassen (1980) that highlight aspects of explanation not so far discussed. According to Achinstein, an explanation is an *illocutionary* act, that is, a speech act like promising or warning that is intended to have a particular effect. Similarly, van Fraassen characterizes an explanation as an answer to a question, and points out that a request for an explanation often occurs in a context that specifies a topic of concern and a set of alternative explanations. To my knowledge, these aspects of explanation have not been investigated in AI, although they would have to be considered in a system fully capable of natural language understanding. The question "Why did *John* phone Mary?" requires a different answer from "Why did John phone *Mary*?" or even "Why did John *phone* Mary?"

Such issues will have to be dealt with as part of a cognitive theory that includes discourse processing, a live area of research in cognitive science that unfortunately has not yet much addressed questions of explanation. Development of a theory of discourse processing will have to be built on top of a theory of language that is built on top of the general theory of cognitive operations that I have been speculating about. Achinstein and van Fraassen have undoubtedly pointed to important features of the *use* of explanation, but their discussions do not eliminate the need to understand the *nature* of explanation in its deductive, statistical, analogical, schematic, and causal strands. The linguistic/pragmatic strand is of relatively minor importance for understanding conceptual revolutions.

One focus of AI research on explanation that I have not yet mentioned concerns the giving of explanations of why programs work the way they do. An expert system may, for example, give a medical diagnosis, but human users will have little confidence in the diagnosis unless some trace of the program's reasoning is presented to them in comprehensible form. Explanation here is not explanation of the symptoms, but explanation of how the program reached the conclusion that it did (see, for example, Buchanan and Shortliffe 1984).

We can expect various of the five major strands of explanation to exhibit themselves in different historical cases. Lavoisier's explanations employing the oxygen theory are too informal to be generally characterized as deductive, although they sometimes do have a quantitative element. Statistical and analogical reasoning do not appear to play any role. There is, however, a schematic component to Lavoisier's explanations, as his basic assumption that air contains oxygen gets applied to different phenomena. Overall, the major strand of Lavoisier's explanations seems to be causal, since the oxygen theory describes a causal mechanism for combustion and calcination. We shall see, however, that in other revolutionary theories different strands of explanation tend to dominate.

5.3.2 Toward an Integrated Cognitive Model of Explanation

For both particular and general reasons, it would be astonishing if any neat definitional analysis of explanation were forthcoming. The particular reasons include the diversity of the six different views of explanation described above, and the fact that all of them seem to capture some relevant aspects of explanation in science and everyday life. The general reasons include the dubiousness of a sharp distinction between definitional, analytic knowledge and theoretical knowledge (Quine 1963), and the view, generally accepted in cognitive psychology and artificial intelligence, that concepts are better characterized by typicality conditions than by necessary and sufficient conditions (Chapter 2). So how can we give a general characterization of explanation?

My answer is that a theory of explanation must be part of a theory of cognitive architecture. I shall briefly review what researchers in AI and cognitive psychology aim to accomplish by developing cognitive architectures, and then locate the place of a theory of explanation within a general theory of cognition. Developing such theories is obviously a gigantic undertaking, and this section is at best preparatory.

COGNITIVE ARCHITECTURES

The notion of a cognitive architecture arose with the early attempt of Newell and Simon (1972) to develop a "general problem solver," but the term "cognitive architecture" apparently originates with Anderson (1983). Anderson (1983, ix) put forward a theory called ACT* (pronounced "act star") as a "theory of cognitive architecture—that is, a theory of the basic principles of operation built into the cognitive system." The aim was to develop a general theory of mind that would embrace all higher-order cognitive operations, including problem solving, memory, learning, and language. ACT* is not a computer program, but a framework within which various programs exist to model different sorts of cognitive operations. ACT*'s representations include production (IF-THEN) rules and a memory consisting of propositions organized into hierarchical networks based on their constituent terms. Its processes include firing of rules and search through memory by spreading activation of nodes in the network. The main direct competitor of ACT* is the SOAR architecture, also developed at Carnegie-Mellon University, which also uses production rules, although in SOAR these fire in parallel (Laird,

Rosenbloom, and Newell 1986). SOAR has successfully modeled many tasks involving problem solving and skill learning, but has not yet been applied to memory or language. Another approach to cognitive architecture is found in the FERMI project, which has at its core a frame-based problem solver (Larkin, Reif, Carbonell, and Gulgiotta 1988).

Recently, a very different approach to the question of a cognitive architecture has arisen on a "PDP" view of cognition as "parallel distributed processing" (Rumelhart and McCelland 1986). The PDP view emphasizes the parallel operation of simple processing units that operate at a lower level than the explicit rules of systems such as ACT* and SOAR. The PDP approach does not yet constitute a cognitive architecture, since no unified scheme for modeling problem solving, memory, and learning has been proposed, although there have been interesting particular models, for example of aspects of language learning. No one has yet proposed a cognitive architecture that takes logical deduction or probabilistic reasoning to be central processes. Cognitive architectures have been proposed by psychologists and psychologically oriented AI researchers, to whom logicist approaches favored by some philosophers (e.g. Pollock 1989) and some AI researchers (e.g. Genesereth and Nilsson 1987) seem implausible.

EXPLANATION IN A COGNITIVE ARCHITECTURE

None of the architectures so far described have said much about explanation, which is, however, a major concern of the system PI (Thagard 1988). I shall now briefly describe how explanation is simulated in PI, incorporating deductive, analogical, and schematic kinds of explanation. A discussion of some of the limitations of PI will point the way to the development of a more successful and richer cognitive architecture.

PI is a rule-based problem solver, in that given a set of starting conditions and a set of goals it can use IF-THEN rules to figure out how to achieve the goals. Use of such rules is quasi-deductive, implicitly involving inferential rules of universal instantiation and modus ponens, except that the rules do not need to be truly universal. Explanation in PI is seen as a kind of problem solving. If E is an explanandum, then the system is set the task of deriving E from other information that it has. However, explanation in PI is not simply deductive, since there are other processes involved than merely deciding what follows from what. PI is also a model of memory, in that not all rules are assumed to be immediately available. Search through memory retrieves past episodes of problem solving or explanation, and analogical reasoning is used to try to convert past solutions into ones for the current case. If a past case is used to solve a current problem, then PI schematizes the two cases into a more general characterization that can be used for future purposes. PI thus incorporates to a limited extent deductive, statistical, analogical, and schematic strands of explanation. The details of how this works are given elsewhere (Thagard 1988). PI falls short, however, of providing a vehicle for a full theory of explanation. My colleagues and I are now building a new system that uses more powerful methods of analogical reasoning and integrates them more fully with rule-based reasoning than PI did (Nelson, Thagard, and Hardy 1991). Schema application will also play a role in the new system.

CAUSALITY

Missing in PI, and in any other AI system I know of, is any subtle appreciation of causal reasoning. There are programs that use causal notions, but none that I know of that could distinguish between cause and effect, or between cause and specious correlation, at a level of sophistication to handle problems like the flagpole example in section 5.3.1. The achievement of such a system does not, however, seem to be beyond the range of AI accomplishments. Whereas implementation of each of the deductive, statistical, analogical, and schematic strands of explanation appears to require special inferential procedures, causal reasoning is best viewed not as a separate module, but as something that is carried out within the context of the other kinds of reasoning. IF-THEN rules can describe causal connections such as "if the ignition key is turned, the car will start." Analogy can be useful in distinguishing real causal relations from accidental ones: given a stock of relations we consider causal, new relations that are similar to established ones are more likely to be seen to be causal than novel ones. In domains familiar enough to have become schematized, general patterns describing causal processes such as gravitational attraction or natural selection can be applied. I conjecture that a cognitive theory of causality can be developed by working out how to implement the many heuristics that we use to detect causality. Discussions of such "cues to causality" have been given by psychologists (Einhorn and Hogarth 1986). Familiar elements on the list, such as covariation, temporal order, contiguity, and similarity, fall far short of providing a theory of causality or an analysis of the concept of cause. But the heuristics provide a reconceptualization of the problems of understanding causality analogous to the reconceptualization of the problem of explanation that I have been urging. The task is no longer to define "cause," but to embed a theory of causal reasoning within a general cognitive theory. In this section, I have only pointed in the general direction; the task of developing such a theory that embraces causality and explanation remains to be accomplished.

Development of a general cognitive theory encompassing explanation will go hand in hand with development of a computational model that can simulate a wide range of cognitive functions. Such a model will help to assuage worries that ECHO relies on an unspecified view of explanation. An integrated cognitive architecture would make possible identification of deductive, schematic, and analogical explanations. ECHO would be an intrinsic part of this system, using parallel constraint satisfaction to evaluate explanatory hypotheses and revise beliefs as new evidence and hypotheses enter the system. The development of a system that gracefully integrates deductive, schematic, and analogical reasoning will contribute to the development of a general cognitive theory of explanation. The aim of this chapter has been much more modest: to identify from current theoretical discussions the various strands of explanation. The historical chapters to come identify these strands in the major conceptual revolutions, and section 10.1.2 summarizes the historical findings.

5.4 SUMMARY

Scientific changes differ in the extent to which new theories incorporate old ones. Scientific revolutions are not strictly cumulative, since new theories involve at least some rejection of the theoretical claims of the theories they replace. The oxygen theory incorporated little of the conceptual system of the phlogiston theory, but was largely cumulative with respect to evidence. Other scientific revolutions vary with respect to the cumulativeness that occurred when new theories were adopted.

There is sufficient continuity in scientific revolutions to justify the claim that adoption of new theories is in general rational. Theory acceptance is not much like religious conversion, and social factors alone do not explain the adoption of new theories. Acquiring a new conceptual system is somewhat similar to learning a second language, but the analogy tends to support the rationality of scientific revolutions.

Explanation has many strands: deductive, statistical, schematic, analogical, causal, and linguistic. A theory of explanation should show how these strands can be smoothly entwined in an integrated cognitive architecture. Although such an architecture has not yet been developed, the prospects for its development are good.

CHAPTER 6

The Darwinian Revolution

CHAPTERS 3 and 4 used the chemical revolution of the eighteenth century to expound a theory of revolutionary conceptual change. We saw that Lavoisier's revolution brought about major changes in kind-relations and part-relations, and that the replacement of phlogiston theory by oxygen theory reflected the greater explanatory coherence of the latter. In the century following Lavoisier's, Charles Darwin brought even more substantial changes to biology than Lavoisier did to chemistry. Not only did he dramatically change the conceptual organization of kinds of organic beings, he changed the principle of organization, making it historical rather than simply based on similarity of organisms. This chapter describes the development of Darwin's ideas and analyzes the changes in kind-relations that he brought about. Darwin's theory of evolution by natural selection was a direct challenge to the dominant scientific view of the first half of the nineteenth century, that species were independently created by God. Darwin's argument against creationism is naturally understood in terms of explanatory coherence.

Have there been biological revolutions besides the one that Darwin produced? Mendelian genetics and molecular biology are the two major developments after Darwin, but a brief discussion will show that they are not revolutionary in the same way, since neither required replacement of previously adopted theories or major conceptual reorganizations. At the end of this chapter, I reject the suggestion, popular among philosophers, that the development and acceptance of scientific concepts is analogous to biological evolution.

6.1 THE DEVELOPMENT OF DARWIN'S THEORY

Since numerous histories of the Darwinian revolution are available (Bowler 1984; Hull 1973; Mayr 1982; Oldroyd 1980; Ruse 1979), there is no need here to attempt a full survey of the development of Darwin's theory. My historical sketch is intended only to provide the background information needed to prepare the way for discussion of cognitive mechanisms involved in the construction of the theory and of conceptual changes that Darwin accomplished.

Darwin conceived the theory of evolution by natural selection in 1838, but did not publish *On the Origin of Species* until 1859. To understand the conceptual changes that his theories produced, we must appreciate the creationist ideas that were dominant at this time. Today, when creationists who deny biological evolution are on the fringes of science, it is difficult to comprehend Darwin's intellectual situation. Virtually everyone, including the leading scientists of the day, believed that God had created species individually. In particular, they believed that God had separately created the human race, endowing our species with unique mental and moral capacities. Although geologists increasingly viewed the earth as having undergone gradual development over millions of years, developments in the biological world were still largely ascribed to divine intervention. Science and religion were not yet divided into separate spheres, and famous scientists wrote treatises proclaiming how the complexity of nature could only be explained by divine creation. William Paley's (1963) natural theology, with its descriptions of wonderfully adapted biological structures such as the eye, impressed many scientists, including the young Darwin. Just as the development of Lavoisier's oxygen theory must be understood against the background of the phlogiston theory, so Darwin's theory of evolution must be seen as a direct challenge to the creationist views of his contemporaries. This challenge was potentially heretical, as well as conceptually revolutionary, which partially explains why Darwin kept his theory to himself and a few trustworthy friends for twenty years.

What was Darwin's theory? He was by no means the originator of the hypothesis that species evolved. For example, Jean Lamarck in 1809 and Robert Chambers in 1844 published books that proclaimed the transmutation of species. Darwin's own grandfather, Erasmus Darwin, had still earlier written a book in verse that advocated what we now call evolution. But Charles Darwin understood evolution very differently from these earlier thinkers, since he saw natural selection as the primary mechanism that produced evolution. According to Lamarck, organisms have both an inherent tendency to increase in complexity and a capacity to inherit acquired characteristics that are useful in particular environments. Darwin did not reject the inheritance of acquired characteristics, a rejection that occurred only with the development of Mendelian genetics early in the twentieth century. But he viewed changes in species that resulted from environmental influences as much less important for biological evolution than the changes that resulted from random variation and natural selection.

Darwin was born in 1809 and graduated, with no particular distinction, from Cambridge University in 1831. From December of 1831 until October of 1836 he served as unpaid naturalist aboard the British Navy ship the *Beagle*, traveling to South America and around the world. Darwin's autobiography describes the intellectual impact of this trip:

> During the voyage of the *Beagle* I had been deeply impressed by discovering in the Pampean formation great fossil animals covered with armour like that on the existing armadillos; secondly, by the manner in which closely allied animals replace one another in proceeding southwards over the continent; and thirdly, by

> the South American character of most of the productions of the Galapagos archipelago, and more especially by the manner in which they differ slightly on each island of the group; none of the islands appearing to be very ancient in a geological sense. It was evident that such facts as these, as well as many others, could only be explained on the supposition that species gradually became modified; and the subject haunted me. (Darwin 1958, 41f.)

But Darwin was not satisfied with available evolutionary hypotheses, because they could not explain such adaptations as the ability of a tree frog to climb trees. He rejected Lamarckian ideas about increasing complexity and the influences of environmental conditions as implausible for explaining adaptation, especially of plants. In 1837 he began a series of notebooks to record reflections on a broad selection of readings relevant to the species question.

The dramatic breakthrough came on September 28, 1838, when a book on population growth by Thomas Malthus suggested to him the mechanism of natural selection. Malthus argued that aid to the poor was of limited value, because the exponential rate of population growth would always outstrip the much slower rate of growth in food supplies. Darwin showed no particular interest in Malthus' political views, but he noticed that the discrepancy between population growth and resources such as food and land would produce intense competition leading to adaptations. Only organisms well adapted to their environments and to reproduction would survive the struggle for existence. Here is the crucial passage from his notebook (Darwin 1987, 375f.):

> One may say there is a force like a hundred thousand wedges trying force into every kind of adapted structure into the gaps of in the economy of Nature, or rather forming gaps by thrusting out weaker ones. The final cause of all this wedgings, must be to sort out proper structure & adapt it to change.

Darwin did not use the term "natural selection" in these notebooks, although it figures prominently in the rough sketch of his theory that he wrote down for his own purposes in 1842. In 1844 he wrote a much longer and more polished description of his views that has much the same structure as his landmark *Origin of Species*. The appearance of that book was prompted by a letter from a naturalist in Malaya, Alfred Wallace, who realized in 1858 that Malthusian ideas about population growth leading to a struggle for existence could explain evolution.

In the *Origin*, Darwin did not address the explosive question of the evolution of humans, but in 1871 he published *The Descent of Man*, which placed our species within the evolutionary framework. Humans, like other animals, have evolved from simpler forms in response to pressures of natural and sexual selection. Wallace never accepted the extension of natural selection to humans, since he thought divine creation was needed to explain our mental and moral faculties.

The next section will describe the dramatic changes in conceptual organization that Darwin's views brought with them. The principal question we must address at the end of this brief historical survey is: what were the principle cognitive mechanisms involved in the generation of Darwin's theory? The two main hypotheses whose origins must be accounted for are:

Species of organic beings have evolved.
Organic beings undergo natural selection.

Because the first of these was not original to Darwin, there is no need to explain how he formed it himself. The passage from his autobiography quoted above shows that he pursued the hypothesis of evolution for reasons of explanatory coherence: evolution could explain the fossil record and other phenomena that he observed on the voyage of the *Beagle*.

Darwin's formation of the hypothesis that selection occurs in nature was clearly explanation-driven: for more than a year he had been trying to find an explanation for why species evolve and adapt. Malthus suggested a "wedging" that would put pressure on species to evolve in adaptive directions. Darwin's discovery of the hypothesis of natural selection is therefore best described as abductive, although it is not so simple as the basic rule-based abduction described in Chapter 3.

A full computational account of the development of Darwinism would also model the origin of the most important new concepts. By far the most important is natural selection, but other concepts invented by Darwin included sexual selection and the struggle for existence. How was natural selection formed? Since the term "natural selection" does not occur in the notebooks, we can say that he created the concept of natural selection after forming the hypothesis that species had evolved because of the Malthusian struggle for existence. Natural selection was clearly not formed by abstraction from examples, nor does it involve coalescence or decomposition (see section 3.1). Rather, the concept arose from processes of conceptual combination producing differentiation from other kinds of selection. Darwin of course possessed the ordinary concept of selection, and he spent much of his time, both in the 1830s and later, studying selection by humans. *Natural* selection provided a differentiation of the concept of selection involving a new kind of selection performed by nature. The computational mechanism that can produce such a differentiation is conceptual combination, which can create new complex concepts out of previous ones. The concept of natural selection can be formed by taking the concept of selection and specifying what does the selection: nature.

The combination of these concepts depended on Darwin's formation of the hypothesis that nature does a kind of selecting, a hypothesis that was suggested to him by Malthus. Previously, Darwin had seen how selection, which he often called "picking," produced breeds of domestic animals adapted to the wishes of their breeders; but he was not able to see how there could be adaptive selection in nature. Adding Malthus to the picture enabled Darwin to see an analogy between artificial and natural selection, because it suggested how nature could select and produce adaptations just as breeders do. Hence Darwin's hypothesis seems to have arisen by a computational process involving both analogy and abduction (Thagard 1988, 62). I expect someday we will have computational models of reasoning and learning that are rich enough to provide a much more detailed account of the conceptual generation displayed in Darwin's notebooks.

6.2 CONCEPTUAL CHANGE IN EVOLUTIONARY THEORY

What was involved in abandoning the creationist explanation of adaptation in favor of Darwin's theory? Much more was required than simply rejecting one set of beliefs and accepting another. Darwin's theory was part of a very different conceptual framework from the creationist one. Not only did Darwin's views involve the introduction of new concepts and the deletion of old ones, they also required conceptual reorganization and the reinterpretation of the kind-hierarchy.

6.2.1 Addition and Deletion of Concepts

We have already seen a short list of new concepts introduced by Darwin: natural selection, struggle for existence, and sexual selection. Although *evolution* was not a new concept, in the context of Darwin's theory it took on a new meaning. Darwin's early writings tended to avoid the term. In the first edition of the *Origin*, Darwin never used the word "evolution," preferring instead to write of "descent with modification"; and the word "evolved" appears only in the final sentence. By the sixth edition, however, Darwin frequently used the term "evolution." His early avoidance of the term is explained by the pre-Darwinian meaning of evolution as a perfection-directed unfolding of potentialities already present in an organism. For Darwin, evolution was based on variation and selection, both of which occurred without any definite direction or guarantee of progress. So the conceptual change involving the term "evolution" is best described as the deletion of central rules concerning evolution—such as that it operates with a goal of increasing perfection—and their replacement by rules describing random variations and natural selection.

What concepts were deleted as the result of the supplanting of creationist explanations by Darwinian ones? Although Darwin himself became less and less religious as he went along, it would be a mistake to say that the concept of God was deleted from his conceptual system. The sixth edition of the *Origin* contains, in its eloquent final paragraph, a reference to the Creator that was not in the first edition. Darwin's theory does not strictly contradict the existence of God, since one can always maintain that it was God who created the universe with laws that eventually resulted in natural selection and biological evolution. Section 6.3 will show, however, that Darwin's theory decisively undermined a powerful argument for God's existence based on the adaptations of organisms.

Although the concept of God can survive in a post-Darwinian nonnatural theology, the concept of special creation was a casualty of the Darwinian revolution. This concept referred to the hypothesized act of a designing Creator producing individual species. Darwin's argument showed that the explanation of adaptation and many other biological phenomena does not require any assumption of special creation.

6.2.2 New Kind-relations

In one of the most understated sentences in the history of science, Darwin remarked in 1859 that by virtue of his theory, "light will be thrown on the origin of man and his history" (Darwin 1964, 488). In the *Origin* he avoided the delicate question of the place of humans in his evolutionary scheme, even though he had been keenly interested in the subject ever since he first started thinking seriously about evolution in the late 1830s. It was only in 1871 that Darwin published *The Descent of Man*, explicitly arguing that humans fall within his evolutionary framework (Darwin 1981). He described the many physical similarities between humans and other organisms that could be explained only by the supposition that humans had evolved like other animals. Even the mental powers and moral sense of humans could be understood in evolutionary terms. From the perspective of the organization of concepts into kind-relations, this was an extremely radical suggestion.

Section 3.1 described branch jumping, the kind of conceptual change that consists of a concept moving from one part of a hierarchy to another. Figure 6.1 shows in schematic form the branch jumping that took place when Darwin brought humans into the evolutionary picture. In the creationist conceptual framework, humans were a special kind of creature, differing decisively from animals in their mental and moral qualities. Darwin argued that humans descended from other forms, so that they should be classified, in accord with principles of evolutionary descent described in the next section, as akin to apes. This shocking reclassification took *human*, a concept that was previously at an independent level in the hierarchy of concepts, and subordinated it to an existing category.

A less striking alteration in kind-relations concerned how animals are

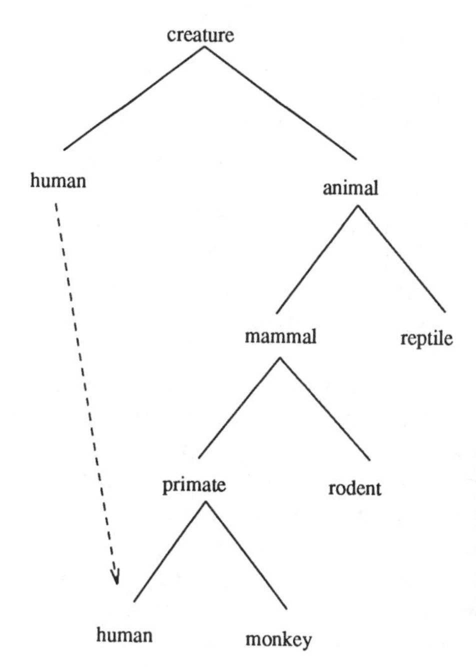

Figure 6.1. Branch jumping in the Darwinian revolution. The straight lines indicate kind-relations. The dotted line shows the movement of the concept *human* in the kind-hierarchy.

grouped into species or varieties. Before Darwin, there was a sharp distinction between species, produced independently by God, and varieties within species that arose because of breeding or natural variation. Darwin broke down the sharpness of the distinction, conceiving of varieties as incipient species. There was thus a collapse of the concept of species and the concept of variety. This conceptual change should not be described as coalescence (section 3.1), since no common term was invented to cover both species and varieties. Darwin wrote (1964, 485): "Hereafter we shall be compelled to acknowledge that the only distinction between species and well-marked varieties is, that the latter are known, or believed, to be connected at the present day by intermediate gradations, whereas species were formerly thus connected." The species/variety distinction breaks down for the fundamental reason that the meaning of *kind* alters fundamentally in the context of the theory of evolution.

6.2.3 Hierarchy Reinterpretation

The concept of *kind* used by Lavoisier and other chemists is no different from the everyday concept that is used in WordNet and artificial intelligence conceptual hierarchies. Approximately, a kind is a group of individuals with similar properties. With Darwin, however, the concept of kind took on a fundamentally historical character, since the classification of organisms into species is to be based on questions of origin in the evolutionary process. Biological classification before Darwin was ahistorical like the ordinary concept of kind. Aristotle, Carolus Linnaeus, and Georges Cuvier devised classifications of organisms based on similarity. But as Oldroyd (1980, 19) notes: "There was no question of the element of time entering Linnaeus's taxonomy. His was an atemporal, or static, representation of the world, based purely on the observations of the external features of living objects. What was directly visible to the naked eye served as the basis for the classification." Cuvier's classification was much more anatomical, sorting organisms on the basis of internal parts rather than external similarity, but it was still based on similarity rather than history.

In contrast, Darwin argued that one advantage of his theory of descent was that it provided a new way of understanding classification. The "grand fact in natural history of the subordination of group under group" is explained by the fact that some groups are descended from others (Darwin 1964, 413). Whereas naturalists had attempted to determine the plan of God's creation by judging the similarity of species, Darwin asserted that the propinquity of descent was the only known cause of the similarity of organic beings. Similarities still play an unavoidable role in an evolutionary approach to classification, since we know so little about the course of evolution; but evolutionary theory including natural selection provides the basis for determining what similarities are important. Similarities in important organs, for example, are more likely to indicate common descent than similarities of mere appearance. In sum: "Our classifications will come to be, as far as they can be so made, genealogies" (Darwin 1964, 486).

This alteration of the fundamental nature of the kind-hierarchy is one of the most revolutionary conceptual changes in the history of science. In Chapter 3, I termed such changes *tree switching*, since the very nature of the hierarchical tree is transformed. Although the methodology of classifying organisms into species remains somewhat controversial, most biologists today accept Darwin's view that classification should be based on the theory of evolution by natural selection and should take commonality of descent as the fundamental principle in constructing a kind-hierarchy (Ridley 1986; Hull 1989). As David Hull (1973, 73) points out, with Darwin species become historical entities, not just groups that share a set of common traits.

According to Hull (1989, 399f.), the hierarchy reinterpretation that comes with the Darwinian revolution is even more drastic than Darwin realized. Hull claims that organisms and species are both historical entities, and as such are connected by part-relations instead of kind-relations. An organism is part of an evolving species, not just a member of that species, and species are parts of higher taxonomic categories. The traditional part-hierarchy that describes genes as parts of cells, cells as parts of organs, and organs as parts of organisms, continues up into what traditionally has been thought of as a kind-hierarchy. If Hull's analysis of species is correct, then the tree switch that was at the core of the Darwinian revolution is the unusual kind that involves a move from kind-relations to part-relations. (In the next chapter, we shall see how plate tectonics involved a move from part-relations to kind-relations.) Moreover, the part-relations in modern evolutionary theory are not simple ones in space alone or in time alone, but simultaneously involve space and time, since species are spatio-temporal entities. They therefore seem similar to the spatio-temporal part-relations that became crucial to modern physics with the introduction of general relativity (see section 8.3).

Even without Hull's interpretation of species, Darwin's theory of evolution brought with it a fundamental change in kind-relations and in the nature of the kind-hierarchy. Now let us consider how explanatory coherence provides a mechanism for moving to this radically new conceptual framework.

6.3 THE EXPLANATORY COHERENCE OF DARWIN'S THEORY

6.3.1 Darwin versus the Creation Hypothesis

When Darwin was a student at Cambridge, the works of William Paley on natural theology were a central part of the curriculum, and Darwin was "charmed and convinced by the long line of argumentation" for divine creation (Darwin 1958, 19). Paley's (1963) argument is naturally interpreted in terms of explanatory coherence. He started with an analogy: If I found a watch, I would naturally infer that the watch must have had a maker, for how else could such a complex and well-designed object have come into existence? Similarly, the many manifestations of complexity in the world, illustrated by chapters on such topics as anatomy and instincts, point to a designer. Figure 6.2 shows the structure of the argument. Just as the watchmaker's work explains the existence of the watch, so does the act of the Creator ex-

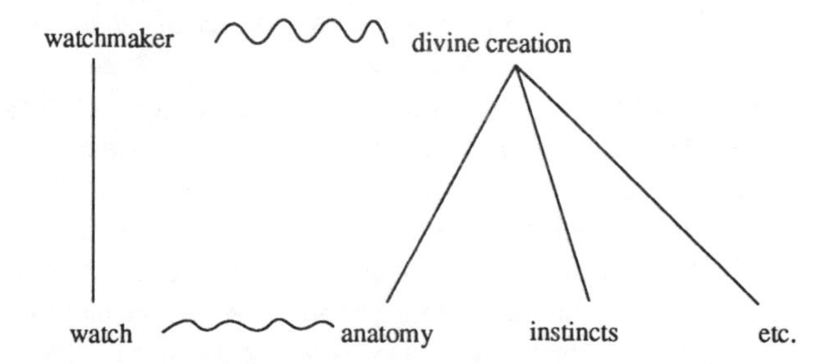

Figure 6.2. The explanatory coherence of Paley's theory of divine creation. The straight lines indicate explanatory relations, and the wavy lines indicate analogies.

plain the complexities of the world. Although this style of reasoning is now restricted to creationists, it was standard in Darwin's day.

Darwin was fully aware of the generally accepted view, held by scientists as well as theologians, that species had been separately created by God. To argue for evolution by natural selection, he had to show that his theory provided a better explanation than did the view of separate creations. That is why there are *dozens* of places in the *Origin* where he explicitly considers the creation theory as an alternative to his view.[1] Thus Darwin's argument is aimed at showing that his theory is better than the alternative creation theory.

6.3.2 ECHO Analysis of Darwin

Darwin described the *Origin of Species* as "one long argument" (1964, 459; 1958, 55). The argument is naturally interpreted as showing that Darwin's theory of evolution has greater explanatory coherence than the creationist theory. The *Origin* starts, not with the evidence for his theory, but with a discussion of variation under domestication. As Darwin explicitly noted, the analogy between artificial and natural selection was an integral part of the justification of his theory (Darwin 1969, vol. 3, 25). After discussing variation under nature, Darwin described the struggle for existence and how it leads to natural selection, which in turn produces evolution (descent with modification). His main hypotheses, that organic beings undergo natural selection and that species of organic beings have evolved, enabled him to explain a host of facts, from the geographical distribution of similar species to the existence of vestigial organs.

Figure 6.3 crudely shows the explanatory coherence of Darwin's theory. The explanatory coherence of the hypothesis that species evolved derives primarily from the host of facts that it helps to explain. The analogy with artificial selection also increases its coherence, as does the fact that evolution is explained by the process of natural selection. Unlike Lavoisier's oxygen theory, Darwin's theory clearly exhibits the importance of analogy and of hypotheses being explained by other hypotheses.

A more detailed ECHO analysis must attend to the explanatory roles of particular hypotheses. In Darwin's explanations, two hypotheses, one about evolution and the other about natural selection, do most of the work (see Table 6.1). But these propositions were not simply cohypotheses, for Darwin also used the latter to explain the former! That is, natural selection explains why species evolve: if populations of animals vary, and natural selection picks out those with features well adapted to particular environments, then

[1] See Darwin (1964), pp. 3, 6, 44, 55, 59, 115, 129, 133, 138, 139, 152, 155, 159, 162, 167, 185, 194, 199, 203, 275, 303, 315, 352, 355, 356, 365, 372, 389, 390, 393, 394, 396, 406, 414, 420, 434, 435, 437, 453, 456, 465, 469, 470, 471, 472, 473, 474, 475, 478, 480, 482, 483, 488. Barrett, Weinshank, and Gottleber (1981) is very useful for tracking such references.

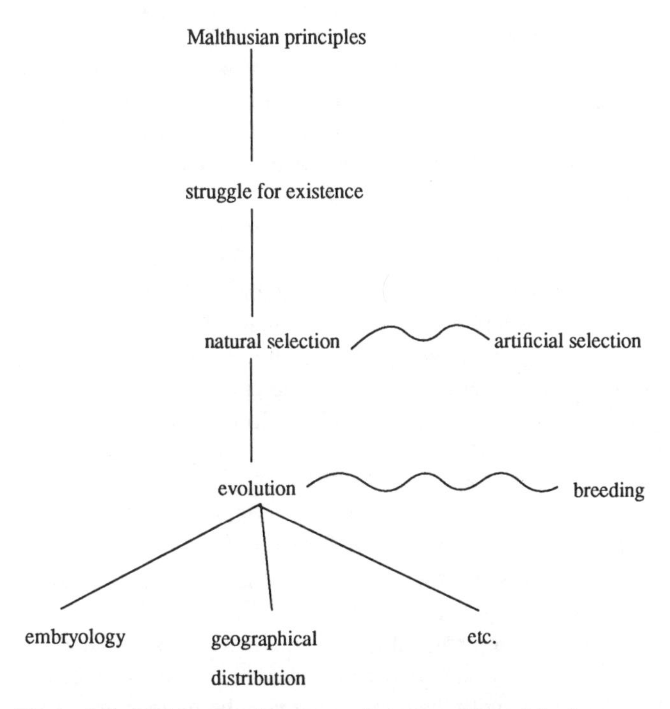

Figure 6.3. Sketch of explanatory coherence of Darwin's theory. The straight lines indicate explanatory relations, and the wavy lines indicate analogies. For a more detailed ECHO analysis, see Figure 6.4.

new species will arise. Moreover, Darwin offered a Malthusian explanation for why natural selection occurs: population growth increases geometrically, as in the series 2, 4, 8, 16; while the available land and food increase at best arithmetically, as in the series 1, 2, 3, 4. Hence population inevitably outstrips available resources, creating intense competition among organisms. Thus Malthusian principles explain why natural selection takes place, which explains why evolution occurs, and natural selection and evolution together explain a host of facts better than does the competing creation hypothesis.

The full picture is even more complicated than this, for Darwin cited the analogy between artificial and natural selection as support for his theory. He contended that, just as farmers are able to develop new breeds of domesticated animals, so natural selection has produced new species. He used this analogy not simply to defend natural selection, but also to help in the explanations of the evidence: particular explanations using natural selection incorporate the analogy with artificial selection. Finally, to complete the picture of explanatory coherence that the Darwin example offers, we must consider the alternative theological explanations that were accepted by even the best scientists before Darwin proposed his theory.

Analysis of *On the Origin of Species* suggests the fifteen evidence statements shown in Table 6.1. E1–E4 occur in Darwin's discussion of objections to his theory, while the others are from the later chapters of the book in which he argues positively for his theory. Table 6.1 also shows Darwin's main hypotheses. DH2 and DH3 are the core of the theory of evolution by natural selection, providing explanations of its main evidence, E5–E15. DH4–DH6 are auxiliary hypotheses that Darwin used in resisting objections based on E1–E3. He considered the objection concerning the absence of transitional forms to be particularly serious, but explained it away by saying that the geological record is so imperfect that we should not expect to find fossil evidence of the many intermediate species his theory requires. Darwin's explanations also use a variety of facts he defended with empirical arguments that would complicate the current picture too much to present here. Hence I shall treat them (DF1–DF7) simply as pieces of evidence that do not need explanatory support. Like E1–E15, DF1–DF7 are marked as data in the input to ECHO (Table 6.2), but their role in the simulation is different since they serve as explainers rather than as evidence to be explained. The creationist opposition frequently mentioned by Darwin is represented by the single hypothesis that species were separately created by God.

Table 6.2 shows the explanation and contradiction statements that ECHO uses to set up its network, which is displayed in Figure 6.4. Notice the hierarchy of explanations, with the high rate of population increase explaining the struggle for existence, which explains natural selection, which explains evolution. Natural selection and evolution together explain many pieces of evidence. The final component of Darwin's argument is the analogy between natural and artificial selection. The wavy lines represent excitatory links based on analogy. Just as breeders' actions explain the development of domestic varieties, so natural selection explains the evolution of species. At another level, Darwin sees an embryological analogy. The embryos of different domestic varieties are quite similar to each other, which is explained by the fact that breeders do not select for properties of embryos. Similarly, nature does not select for most properties of embyros, which explains the many similarities of embryos of different species.

DH3, that species evolved, and CH1, that species were separately created by God, are specified as contradictory, so ECHO sets up an inhibitory link between them. In addition, on the basis of the Principle of Competition, ECHO sets up inhibitory links as follows:

DH2 competes with CH1 because of (E1 E2 E3).
DH4 competes with CH1 because of (E1).
DH5 competes with CH1 because of (E2).
DH6 competes with CH1 because of (E3).

Figure 6.4 shows the inhibitory links that these competitions produce.

Table 6.1
Propositions for Darwin Example

Darwin's evidence:	
E1	The fossil record contains few transitional forms.
E2	Animals have complex organs.
E3	Animals have instincts.
E4	Species when crossed become sterile.
E5	Species become extinct.
E6	Once extinct, species do not reappear.
E7	Forms of life change almost simultaneously around the world.
E8	Extinct species are similar to each other and to living forms.
E9	Barriers separate similar species.
E10	Related species are concentrated in the same areas.
E12	Species show systematic affinities.
E13	Different species share similar morphology.
E14	The embryos of different species are similar.
E15	Animals have rudimentary and atrophied organs.
Darwin's main hypotheses:	
DH1	Organic beings are in a struggle for existence.
DH2	Organic beings undergo natural selection.
DH3	Species of organic beings have evolved.
Darwin's auxiliary hypotheses:	
DH4	The geological record is very imperfect.
DH5	There are transitional forms of complex organs.
DH6	Mental qualities vary and are inherited.
Darwin's facts:	
DF1	Domestic animals undergo variation.
DF2	Breeders select desired features of animals.
DF3	Domestic varieties are developed.
DF4	Organic beings in nature undergo variation.
DF5	Organic beings increase in population at a high rate.
DF6	The sustenance available to organic beings does not increase at a high rate.
DF7	Embryos of different domestic varieties are similar.
Creationist hypothesis:	
CH1	Species were separately created by God.

Table 6.2
Explanations and Contradiction for Darwin Example

Darwin's explanations:

(a) of natural selection and evolution.
(explain (DF5 DF6) DH1)
(explain (DH1 DF4) DH2)
(explain (DH2) DH3)

(b) of potential counter-evidence
(explain (DH2 DH3 DH4) E1)
(explain (DH2 DH3 DH5) E2)
(explain (DH2 DH3 DH6) E3)

(c) of diverse evidence
(explain (DH2) E5)
(explain (DH2 DH3) E6)
(explain (DH2 DH3) E7)
(explain (DH2 DH3) E8)
(explain (DH2 DH3) E9)
(explain (DH2 DH3) E10)
(explain (DH2 DH3) E12)
(explain (DH2 DH3) E13)
(explain (DH2 DH3) E14)
(explain (DH2 DH3) E15)

Darwin's analogies:

(explain (DF2) DF3)
(explain (DF2) DF7)
(analogous (DF2 DH2) (DF3 DH3))
(analogous (DF2 DH2) (DF7 E14))

Creationist explanations:

(explain (CH1) E1)
(explain (CH1) E2)
(explain (CH1) E3)
(explain (CH1) E4)

Data:

(data (E1 E2 E3 E4 E5 E6 E7 E8 E9 E10 E11 E12 E13 E14 E15))
(data (DF1 DF2 DF3 DF4 DF5 DF6 DF7))

Contradiction:

(contradict CH1 DH3)

Darwin's discussion of objections suggests that he thought that creationism can naturally explain the absence of transitional forms and the existence of complex organs and instincts. Darwin's argument was challenged in many ways, but on his own view of the relevant explanatory relations, at least, the theory of evolution by natural selection is far more coherent than the creation hypothesis. Creationists, of course, would marshal different arguments.

Figure 6.5 shows the connectivity of DH3, hypothesizing the evolution of species, in more detail, indicating the strengths of the excitatory and inhibitory links established by ECHO and the final activations of the various units linked to DH3. Running ECHO to adjust the network produces the expected result: Darwin's hypotheses are all activated while the creation hypothesis is

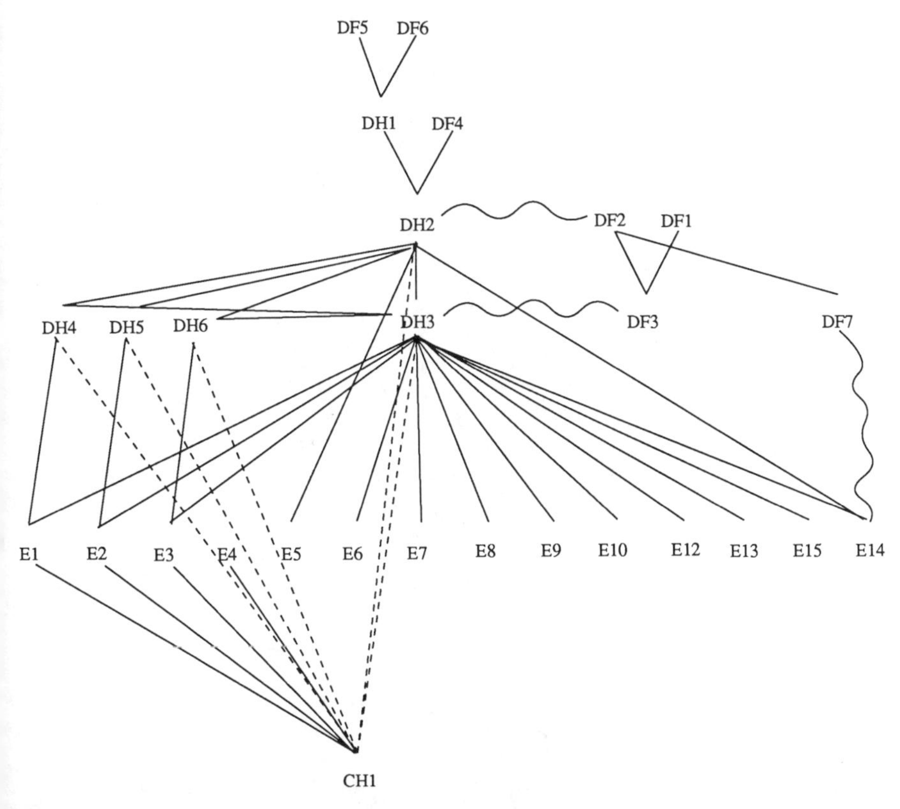

Figure 6.4. Network produced by ECHO from input in Tables 6.1 and 6.2. E1–E15 are evidence units. DH2 represents natural selection, and DH3 represents evolution of species. These defeat CH1, which represents the hypothesis that species were independently created. Solid lines are excitatory links, while dotted lines are inhibitory.

deactivated. DH3 accrues activation in three ways. It gains activation from above, from being explained by natural selection, which is derived from the struggle for existence, and from below, by virtue of the many pieces of evidence it helps to explain. In addition, it receives activation by virtue of the sideways analogy-based links with explanations using artificial selection. Figure 6.6 graphs the activation histories of most of the units over the 53 cycles it takes them to settle. Note that the creationist hypothesis CH1 initially gets activation by virtue of what it explains, but is driven down by the rise of DH3, which contradicts it.

What is the nature of the explanations used by Darwin? His derivation of natural selection from the struggle for existence is roughly deductive, although much less rigorous than mathematical proof. As Kitcher (1981) suggested, Darwin's explanations have a major schematic component, since he

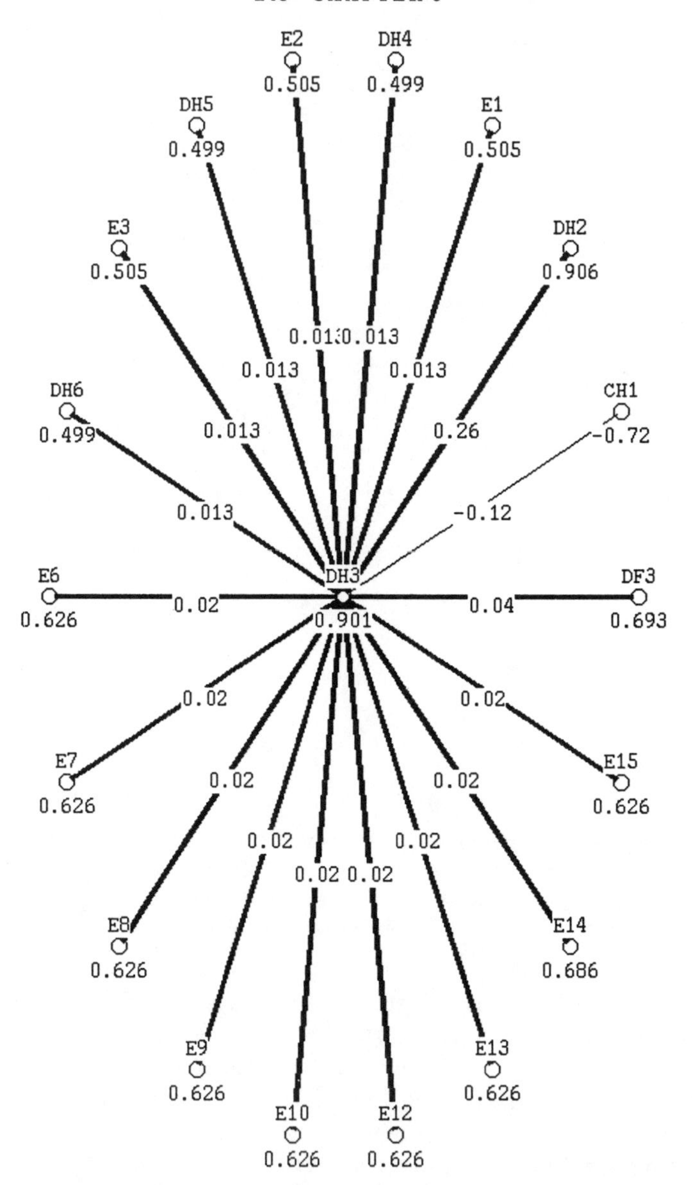

Figure 6.5. Connectivity of unit DH3 in the Darwin network. The numbers under the units are their activation values after the unit has settled. Thick lines indicate excitatory links, while the thin line indicates an inhibitory link. Numbers on the lines indicate the weights on the links.

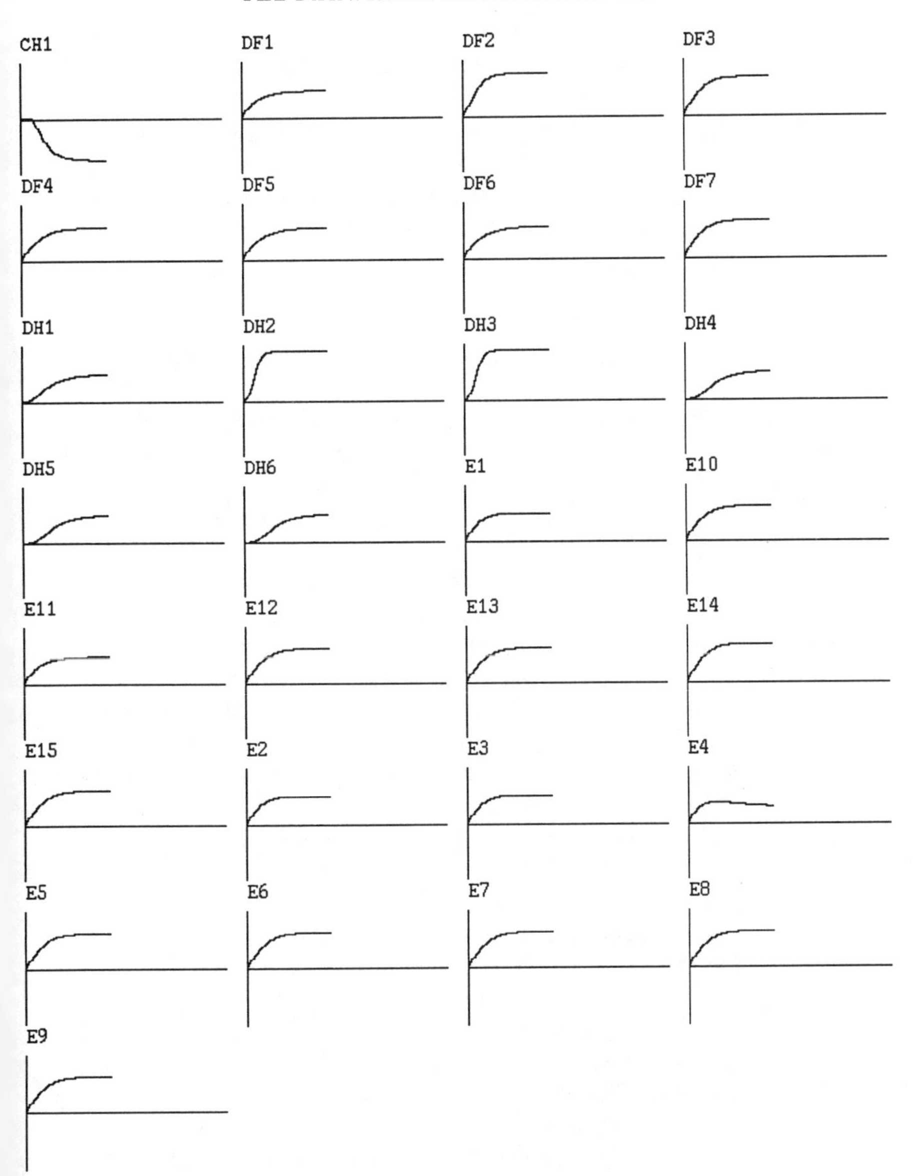

Figure 6.6. Activation history of the Darwin network. Each graph shows the activation of a unit over 53 cycles of updating, on a scale of –1 to 1, with the horizontal line indicating the initial activation of 0.

applied the same pattern of explanation involving evolution by natural selection to diverse facts. In addition, we have also seen that analogy played a role in some of Darwin's explanations, and his discussion of variability was implicitly statistical. Natural selection is undoubtedly a causal mechanism: we can interpret the ECHO analysis as saying that variability and the struggle for existence causes natural selection, and natural selection causes evolution of species. Thus Darwin's explanations involve all the main strands of explanation identified in section 5.3.

6.3.3 Relation of Darwin's Theory to Its Predecessor

Section 5.1 described various relations that a theory can have to one that it replaces. Darwin's theory of evolution by natural selection obviously involved a massive rejection of the theory of divine creation. While not strictly abandoning the concept of God, it rejected the proposition that God played a direct role in producing species. Moreover, it removed the traditional special place held by the concept *human* in the classification of animals, and fundamentally altered the nature of the kind-hicrarchy and the procedures of classification. Divine intervention ceased to be among the explanatory principles of biology, so the relation between Darwin's theory and its creationist predecessor is best described as *supplantation*. Certainly, Darwin did not incorporate the divine creation theory within his own, but he did not ignore it either. The many comparisons of the two theories in the *Origin* show the premium he placed on showing the superiority of evolution to creation in explaining the biological facts. So Darwin did not disregard the alternative theory any more than he incorporated it.

Although there was little theory cumulation from Paley to Darwin, there was substantial evidence cumulation, since Darwin tried in his discussion of objections to show how his theory could explain the facts that were most strongly explained by divine creation. His explanations of the gaps in the fossil record and of phenomena such as instincts were somewhat ad hoc, but he clearly did not abandon concern with the evidence that creationists had cited for their theory: for both Darwin and Paley, adaptation was the key fact to be explained. Method also seems to be cumulative between Paley and Darwin as far as argument goes, since the explanatory structure of Paley is similar to that of Darwin, in that both use analogy and claims about what their theories explain. Of course, Darwin opened up a fertile new way of thinking about species in terms of populations described in statistical terms, and made many other conceptual advances. But the Darwinian revolution had a substantial degree of evidence and method cumulation, even as the old theory was dramatically abandoned.

6.3.4 Other Interpretations of Darwin's Argument

Doren Recker (1987) and Elisabeth Lloyd (1983) have challenged my earlier claim that Darwin's argument in *Origin of Species* should be understood as defending an inference to the best explanation (Thagard 1978). Lloyd, in keeping with a semantic conception of theories, claims that Darwin was trying to show the empirical adequacy of his theory. Recker sees in the *Origin* three strategies for showing the *causal efficacy* of Darwin's theory. The first involves showing that natural selection is a *vera causa* (true cause), the second responds to objections, and the third concerns the explanatory power of the theory.

LLOYD ON EMPIRICAL ADEQUACY

According to Lloyd, Darwin should be understood as defending the empirical adequacy of his theory. She reinterprets his numerous statements concerning the explanatory power of his theory as merely claiming that there is a fit between theory and data. This account, which owes much to the views of van Fraassen (1980), seems to differ from an inference-to-the-best-explanation account on two dimensions. First, the latter account is explicitly comparative, advocating the acceptance of the *best* of available explanations. Second, the latter is much more consonant with a realist view of scientific theories, since the best explanation is typically inferred to be *true*. We have already seen that Darwin's argument was explicitly comparative, and there is also ample evidence that he was a realist.

Darwin's statements in the final edition of the *Origin* and in other writings show that he took the explanatory power of his theory to be evidence of its *truth*:

> It can hardly be supposed that a false theory would explain, in so satisfactory a manner as does the theory of natural selection, the several large classes of facts above specified. It has recently been objected that this is an unsafe method of arguing; but it is a method used in judging of the common events of life, and has often been used by the greatest natural philosophers. (Darwin 1962, 476)

> I must freely confess, the difficulties and objections are terrific; but I cannot believe that a false theory would explain, as it seems to me it does explain, so many classes of facts. (Darwin 1903, vol. 1, 455)

> But whether it [the assumption of preservation of variations] is true, we can judge only by seeing how far the hypothesis accords with and explains the general phenomena of nature. (Darwin 1962, 94)

Thus Darwin was not merely trying to argue that his theory was empirically adequate: he was maintaining that it should be accepted as true because it provided a better explanation than the alternative creation theory.

RECKER ON CAUSAL EFFICACY

The three strategies that Recker sees Darwin using to support the causal efficacy of evolution by natural selection correspond to three parts of Darwin's book. The first part lays out the basic ideas of natural selection, the second discusses objections to the theory, and the third shows its explanatory power. What Recker sees as the third part of Darwin's strategy is obviously fully compatible with the view that Darwin was using inference to the best explanation. I shall show that the first two parts of the argument also are best understood as contributing to the explanatory coherence of his theory.

Darwin had been led to suspect that species evolved early on, but had become convinced of it only when he realized that natural selection could explain *how* species had evolved. Moreover, he derived natural selection from the struggle for existence which in turn followed from Malthusian principles about the geometrical rate of increase of population. Thus population pressures explain the struggle for existence which explains natural selection which explains the evolution of species. Moreover, natural selection and species evolution together explain the host of facts that Darwin discusses in later chapters of the *Origin.* We saw how all these explanations, as well as Darwin's use of analogy, can contribute to explanatory coherence. Thus the first part of Recker's analysis fits well within my explanatory coherence account.

Now let us look at Darwin's responses to objections. According to Recker, the chapters in which Darwin discusses the objections to his theory constitute a separate part of his argument for the causal efficacy of natural selection. But close examination, bearing in mind the comparative nature of Darwin's argument, shows that they are in fact directly concerned with whether his theory is the best explanation of the evidence.

The four major objections that Darwin considers are: the fossil record contains few transitional forms, animals have complex organs, animals have instincts, and animals crossed from different species are usually sterile. Not only are these facts difficult to explain on Darwin's theory, they fit well with the alternative creation hypothesis. Separate creation would explain why few transitional forms have been observed, why animals were provided with complex organs and instincts, and why animal species separately created by God remain separate. Darwin's replies to these objections involve his postulation of auxiliary hypotheses to explain these facts, which he realizes are more simply explained on the rival view. For example, he explains the absence of fossils that show transitional forms by arguing that the geological record is highly imperfect, and he sketches explanations of how instincts and complex organs might have evolved by natural selection. Thus the discussion of objections is directly relevant to the explanatory power of his theory compared to the alternative creation hypothesis. Contrary to Recker's opinion, it is not a separate strategy concerned with causal efficacy. Explanatory coherence theory is all one needs to appreciate Darwin's argument.

6.3.5 The Reception of Darwin's Argument

For Darwin, the hypotheses of evolution and natural selection were both key parts of his theory, but in the decades following the publication of the *Origin* they received very different receptions. By 1870, most scientists had accepted evolution, a major turnaround from before Darwin's work. But natural selection remained highly controversial, and did not become well established as a principle of biology until the development of the synthesis of evolutionary biology with genetics in the 1930s. ECHO has given an adequate account of the coherence of Darwin's theory *for him*, but it should also be able to account for why other thinkers found natural selection much less palatable than evolution. I see two major reasons why natural selection might appear to Darwin's contemporaries as less acceptable than evolution, both based on incoherencies between it and strongly held propositions. Even religious people could give up the hypothesis of special creation, holding on to the view that God had designed evolution. But natural selection provided a mechanism based on random variations that challenged the easy retreat of the theological view. In Figure 6.4, I showed special creation as contradicting evolution, but Figure 6.7 shows another theological hypothesis, divine design, contradicting natural selection. Divine design is much less specific than special creation and is much more easily made compatible with scientific facts about extinction of species and their appearance at different times. Special creation could be abandoned while retaining the more deeply rooted hypothesis that God played at least some role in designing the species that appeared intermittently.

Another incoherency for natural selection was due to an influential argument of Lord Kelvin's. Using the best physical ideas of the day, he calculated the age of the earth to be much less than would have been necessary for the inherently slow process of natural selection to have had time to bring about the development of so many different species. All Darwin could do was hope that Kelvin had miscalculated somewhere. As it turned out, Kelvin was shown to be mistaken decades later, when the discovery of radioactivity suggested a heat source in the earth that could keep it from cooling as rapidly as he had thought. In the meantime, the hypothesis of natural selection could be seen as incoherent with well established principles of physics. So natural selection fares worse in a calculation of explanatory coherence than does evolutionary theory.

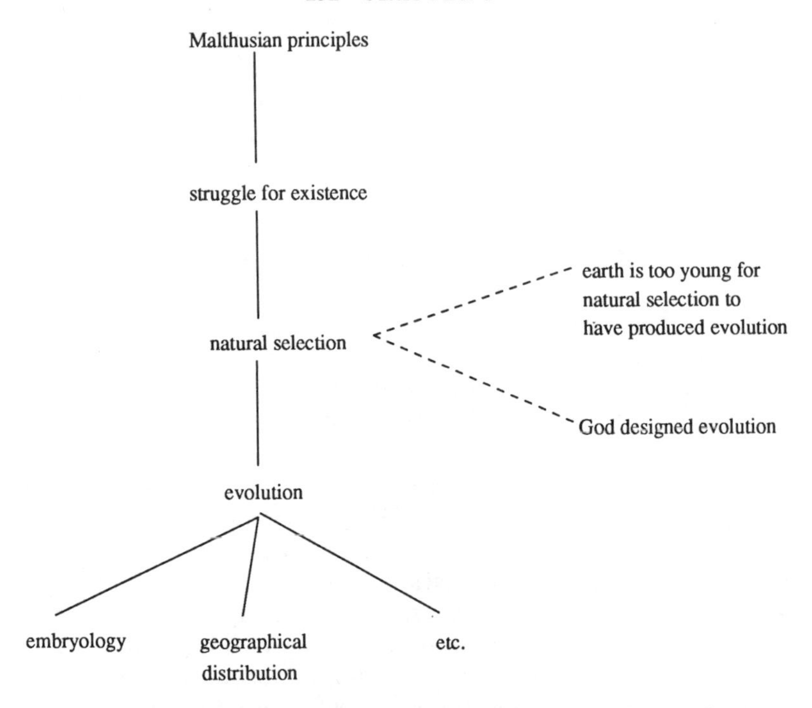

Figure 6.7. Incoherence of natural selection with other beliefs. Straight lines indicate the explanatory relations of Darwin's theory, as in Figure 6.3. Dotted lines show the contradictions between natural selection and other widely held views of the time.

There is an important lesson here about explanatory coherence. The coherence of a theory is not just a matter of how it competes head to head against alternative explanations of a body of evidence. In principle, coherence involves relations to everything else one believes. Hence a hypothesis with good explanatory credentials might nevertheless be rejected if it contradicts other well established parts of science.

6.4 OTHER REVOLUTIONS IN BIOLOGY?

Historians and biologists are agreed that the development of Darwin's theory of evolution was the most revolutionary episode in biology to date, but have there been other revolutions in biology? According to Cohen (1985), William Harvey initiated a revolution in the life sciences when he proposed his theory of the circulation of the blood. On my account of conceptual revolutions, the central features of radical conceptual change are:

1. replacement of one theory by another that has greater explanatory coherence; and
2. conceptual reorganization substantially affecting kind-relations and/or part-relations.

The first feature is clearly present in Harvey. He explicitly argued against Galen's view, dominant for many hundreds of years, that the heart propelled blood out to the body where nutrients were absorbed, without any circulation. An ECHO analysis of Harvey's theory should be easy to do, describing how he used the hypothesis of circulation to account for the functions of the venous and heart valves and for quantitative facts concerning the volume of blood flow. As Harvey (1962, 139) summarizes:

> All these appearances, and many others, to be noted in the course of dissection, if rightly weighed, seem clearly to illustrate and fully confirm the truth contended for throughout these pages, and at the same time to oppose the vulgar opinion; for it would be very difficult to explain in any other way to what purpose all is constructed and arranged as we have seen it to be.

Working out in detail the explanatory relations of Harvey's and Galen's theories is left to other researchers.

Harvey's theory also brought with it an interesting conceptual reorganization, particularly respecting part-relations. According to Cohen (1985), whereas Galen viewed the liver and veins as a system wholly distinct from the system of heart and arteries, Harvey initiated the medical view of all these as a single system. Harvey is considered to be the founder of modern medicine and it would be interesting to analyze the kind-relations and part-relations in his framework and to contrast them with the concepts of Galen. The adoption of Harvey's theory thus appears to be a full-blown conceptual revolution.

The two major developments in biology since Darwin are genetics and molecular biology. Although both of these are scientific developments of enormous importance, they do not count as revolutionary according to the account of conceptual change developed here. Before genetic theory there was no established theory of inheritance. Darwin had proposed his theory of pangenesis, according to which all parts of the body contribute genetic material to the reproductive organs, but even he did not find it very convincing. Thus genetics did not involve the overturning of a previous view. The grand synthesis of genetics with Darwin's theory of evolution by natural selection in the 1930s increased the coherence of Darwin's theory by explaining how inheritance takes place. Moreover, although genetics involved interesting new part-relations (Darden and Rada 1988), it did not require revision of previous part-relations. Genes are parts of chromosomes. Decomposition is an important strategy for introducing new parts, but it generally produces incremental changes in the conceptual structure, not revolutionary ones. Similarly, introducing quarks in subatomic physics added a new layer to the part-hierarchy, but did not require reorganizing higher layers (see section 8.5). The development of the concept of a gene is a fascinating and important topic (Darden 1991), but it is not part of a conceptual revolution.

Molecular biology has opened up wonderful new avenues of research, but has not required overturning the established theories of evolution and genetics. Genetic theory is improved dramatically by the understanding of genes as composed of DNA, but the discovery of DNA and later developments in molecular biology primarily added part-relations rather than revising previously established ones. The advent of molecular biology did not require any notable abandonments of theory, evidence, or method.

6.5 CONCEPTUAL EVOLUTION?

Darwin provided us with a theory of evolution by natural selection that has great explanatory coherence in the biological sphere. Not surprisingly, attempts have been made to apply theories of the same type in other spheres (Darden and Cain 1989). Philosophers of science have appealed to evolutionary models to account for the development of scientific knowledge (Toulmin 1972; Campbell 1988). Evolutionary epistemology is problematic because of the numerous differences between conceptual development and species development. For example, concept generation in science is generally part of the problem solving process and is data-driven, explanation-driven, or coherence-driven. Organisms, on the other hand, do not generate mutations in order to overcome environmental problems. In my earlier book, I argued that a computational perspective highlights the inadequacy of using biological evolution as a model of scientific change (Thagard 1988, ch. 6). I shall not repeat those arguments here, but want to evaluate the recent sophisticated attempt by Hull (1989) to understand conceptual change by analogy with biological evolution. Especially for conceptual revolutions, the cognitive/computational approach advocated here has numerous advantages over the biological approach.

Unlike many evolutionary epistemologists, Hull does not simply draw analogies between biological and conceptual evolution. Instead, he develops a set of high-level categories intended to apply to both. The key notions are (Hull 1989, 408f.):

replicator—an entity that passes on its structure largely intact in successive replications.

interactor—an entity that interacts as a cohesive whole with its environment in such a way that this interaction *causes* replication to be differential.

selection—a process in which the differential extinction and proliferation of interactors *cause* the differential perpetuation of the relevant replicators.

In biological evolution, genes are the paradigmatic replicators, and organisms are the paradigmatic interactors. Selection is a process in which the differential fitness of organisms causes the differential perpetuation of their genes.

In science, the replicators are "beliefs about the goals of science, proper ways to go about realizing these goals, problems and their possible solutions, modes of representation, accumulated data, and so on" (Hull 1989, 434). The primary interactors are scientists who generate new replicators, including concepts. The process of selection in science involves competition among scientists to have their views accepted by other scientists.

As far as it goes, this analysis of the process of science in terms of replicators, interactors, and selection is unexceptionable, and it is useful for characterizing the social side of science, which often has different scientists competing with each other. But the biologically inspired analysis is inherently limited, since it cannot describe in any detail the processes by which new concepts and hypotheses are formed and evaluated, nor can it describe the organization of concepts. These are fundamentally psychological questions that the biological analysis cannot address. What is the organization of replicators? Scientific concepts are organized by kind-relations and part-relations, while hypotheses are organized by relations of explanatory coherence. Using cognitive notions rather than biological abstractions, we are able to characterize many different kinds of conceptual change.

Evolutionary epistemology usually assumes the orthodox interpretation of biological evolution according to which species evolve as the result of long periods of gradual changes. Conceptual revolutions, however, are better understood by comparison with the controversial theory of *punctuated equilibrium*, according to which biological evolution is not a steady, gradual process of variation and selection (Eldredge 1985). Instead, long periods with little change are interrupted by dramatic periods when species emerge or become extinct at a much greater rate. I shall not address the biological merits of this theory, but there can be no question that science is such a punctuated process. Scientific revolutions are rare, with only seven or eight major ones in three fields in a period of five hundred years. Comprehensive explanatory theories like Darwin's and Newton's are exceptional, and the overturning of them once they are established is rarer still. Hull's biologically inspired analysis is useful for sociological description, but it cannot contribute to psychological understanding of the major conceptual changes that attend scientific revolutions.

Defenders of evolutionary epistemology might claim that I have merely been proposing a different analogy than them, modeling conceptual development in computational rather than in biological terms. But if the strong claims of cognitive science are valid, thinking is not just *like* computation, it *is* computation. Currently, we use computers dissimilar to the brain to develop cognitive theories, but we can aim to have computers that think much in the same way that people do. For computational philosophy of science, this aim involves the ambition of building programs that construct concepts and hypotheses in approximately the way that scientists do. The ambition is far from being realized, but the philosophical arguments against it are unconvincing, and many promising research directions are currently being pursued. Ultimately, the choice between the research program of evolutionary epistemology and the cognitive/computational research program should be made according to which program generates theories with greater explanatory coherence. I hope that impartial observers will view the current theory of conceptual change as evidence that the computational/cognitive approach can help to generate richer models of science than biological analogies.

6.6 SUMMARY

The development of Darwin's theory of evolution by natural selection was driven by the need to explain several puzzling biological facts. Using Malthusian principles, Darwin abductively formed the hypothesis that a struggle for existence had led to species diversity, and he created the combined concept of natural selection. With respect to the previously dominant creationist theory, Darwin's theory involved several kinds of conceptual change. It introduced novel concepts like natural selection and changed the kind-hierarchy by reclassifying humans as a kind of primate. Moreover, Darwin redefined the kind-hierarchy, since historical lineage rather than similar features became the key principle of classification. Darwin's long argument in the *Origin of Species* justified the adoption of the new conceptual system by showing that his theory had much greater explanatory coherence than the opposing creationist theory. He could explain more facts, and in addition gained coherence from the application of Malthusian principles and the analogy between natural and artificial selection. Darwin's theory supplanted the creationist theory, whose concepts and principles were largely abandoned, although the evidence and method of arguing for it were largely preserved in the Darwinian system.

CHAPTER 7

The Geological Revolution

THE twentieth century has seen three major conceptual revolutions, associated with the theories of relativity, quantum mechanics, and plate tectonics. The physics cases must wait until the next chapter, for I want first to consider the most recent scientific revolution, in geology.[1] During the 1960s, the theory of plate tectonics became the dominant framework for understanding geological phenomena. This chapter is an attempt to interpret the development of geological ideas in this century in terms of conceptual change and explanatory coherence. It traces the history of the hypothesis of continental drift that preceded modern plate tectonics, and explains why the hypothesis was not generally accepted until it became reformulated as part of plate tectonics. The adoption of plate tectonics required major reorganization in the conceptual hierarchies that included concepts such as *continent* and *seafloor*.

7.1 THE CONCEPTUAL DEVELOPMENT OF PLATE TECTONICS

7.1.1 Historical Sketch

The history of ideas about continental drift and plate tectonics has been told in detail elsewhere (Hallam 1973; Glen 1975; Menard 1986), so only a brief outline is needed here. In 1915 Alfred Wegener, a German meteorologist, published a book in which he brought forward a mass of geological and paleontological evidence supporting the hypothesis that the continents had once been joined together and drifted apart. Although Wegener's book went through four German editions and was translated into English and French, his theory of continental drift was not generally accepted. The dominant global theory for explaining geological phenomena at the time was based on the hypothesis that the earth was cooling and therefore contracting, with contraction causing such phenomena as mountain building.

Although continental drift had some supporters during the following decades, it remained on the fringe of geology until the early 1960s. During the 1950s, the U.S. Navy sponsored expeditions to map the seafloor in both the Atlantic and Pacific oceans. These expeditions brought to light many new

[1] This chapter is based on articles written with Gregory Nowak.

phenomena, in particular information about mid-ocean ridges. In 1960, Harry Hess (1962) and, independently, Robert Dietz (1961) proposed the hypothesis of *seafloor spreading*, according to which the ridges were caused by an upward flow of mantle material which then spreads out, creating new seafloor and carrying the continents away from each other. This is a very different mechanism from the one Wegener proposed, as Hess (1962, 609) takes pains to make clear: "The continents do not plow through the oceanic crust impelled by unknown forces; rather they ride passively on mantle material as it comes to the surface at the crest of the ridge and then moves laterally away from it."

In 1963 F. Vine and D. Matthews used the hypothesis of seafloor spreading to predict magnetic anomalies that could be expected to appear parallel to oceanic ridges if seafloor spreading occurs. The initial evidence collected by ocean surveys did not clearly confirm the Vine-Matthews prediction, but in 1965 the survey ship *Eltanin* found striking magnetic patterns that fit perfectly with what Vine and Matthews had predicted. The new view of the continents and seafloor was elaborated by J. Tuzo Wilson's (1965) concept of a transform fault, his proposal that the crust consisted of a set of large, rigid plates, and Jason Morgan's (1968) development of a mathematical framework for what has come to be called plate tectonics. By the end of the 1960s, plate tectonics provided the generally (but not universally) accepted framework for understanding diverse geological phenomena. More historical details are provided below.

What would be required for a full computational understanding of these important developments? At the least, we need a set of mechanisms sufficient for generating the key concepts and hypotheses. In addition, we need an account of the conceptual structures and transformations that took place as part of the geological revolution. Finally, we need algorithms for evaluating continental drift and seafloor spreading; to be historically adequate, a model incorporating these algorithms will have to account for the rejection of Wegener's theory and the acceptance of the 1960s view.

7.1.2 The Discovery of Continental Drift

The theses on conceptual revolutions in Chapter 1 conjectured that theoretical concepts and hypotheses arise primarily by the mechanism of conceptual combination and abduction. I shall now sketch how these mechanisms, as implemented in the system PI (section 3.4 above; Thagard 1988), can be applied to the early development of continental drift. Although I shall describe an actual run of PI, I stop well short of claiming to simulate the discovery of continental drift, since PI is given only the minimal amount of information needed to operate the relevant mechanisms.

We have a direct quote from Wegener that reports how he first came to think of continental drift: "The first concept of continental drift came to me as far back as 1910, when considering the map of the world, under the direct impression produced by the congruence of the coastlines on either side of the Atlantic" (Wegener 1966, 1). Wegener noticed the large extent to which the coastline of Africa can be fitted to the coastline of South America. The hypothesis that the continents were once joined together and later drifted apart provides an explanation for the degree of fit. Wegener later collected a vast amount of paleontological and geological evidence in favor of continental drift, but the congruence of the coastlines was the stimulus for the initial discovery.

A crude version of this discovery has been implemented in the processing system PI, which performs various kinds of induction in the context of problem solving. Explanation is understood as a kind of problem solving where the facts to be explained are treated as goals to be reached, and hypotheses can be generated to help provide the desired explanations. Problem solving proceeds by the spreading activation of concepts and the firing of rules. To get it started, PI is given the goals of explaining two propositions whose English representations are:

E1. South America and Africa fit together.
E2. South America and Africa are apart.

The concepts *fit together* and *are apart* have associated rules that can potentially provide explanations of these jointly puzzling facts.

R1. If *z* split into *x* and *y*, then *x* and *y* fit together.
R2. If *x* and *y* were together and drift, then *x* and *y* are apart.

The abductive inference using E1 and R1 is more complicated than the kind described in section 3.4.2, since it requires positing that there was a *z* that split into South America and Africa: PI does abduction with *n*-place relations like "split" as well as simple predicates like "F". In a kind of inference called *existential abduction*, PI makes this inference, concluding that some *z* (what Wegener called Pangaea) split into South America and Africa; this hypothesis explains why they now fit together. Similarly, R2 can be used to explain E2 by forming the hypotheses that South America and Africa drift. PI does not need to abduce that South America and Africa were together, since it deduces this fact using an additional rule:

R3. If *z* split into *x* and *y*, then *x* and *y* were together.

Having abduced that South America and Africa drift, PI can generalize that continents drift because it was also informed that they are continents. In PI, generalization of "all A are B" occurs when there are enough A's that are B's to justify the generalization given background information about variability (see Holland et al. 1986, ch. 8). The continental drift simulation would require background information about how things like continents vary with respect to properties like drifting. Variability is normally calculated on the basis of additional knowledge in the data base. Rather than build up the data base, I merely told the program that variability is low, so two instances (and the absence of counterexamples) suffice for the generalization that continents drift. This inference is triggered by the simultaneous activation of the propositions that South America and Africa are continents and the propositions that each drifts.

Similarly triggered in PI is the attempt to combine old concepts into new ones. Here, PI joins CONTINENT and DRIFT into CONTINENT-DRIFT, which can be understood as the concept of a continent that drifts. This is still short of the full concept of continental drift, which refers to a process rather than instances of drifting continents, but it is a start. At the end of its short run, PI has thus formed several hypotheses about South America and Africa and the general rule that continents drift, as well as the new combined concept.

This simulation has many limitations. Unlike Wegener, PI is not capable of noticing for itself that the fit of South America with Africa needs explaining. PI was provided with only limited information: it lacks understanding of the various geological processes involved. A richer representation scheme may be needed, perhaps the qualitative process system of Forbus (1985). I have not provided detailed description of PI's mechanisms of problem solving, abduction, and conceptual combination, since these are available elsewhere (Thagard 1988). Nevertheless, this sketch indicates the relevance of those mechanisms to the problem of explaining the discovery of continental drift.

7.2 CHANGES IN CONCEPTUAL STRUCTURE

7.2.1 Hierarchy Transformation

There were two major kinds of conceptual change in the geological revolution. First, a host of new concepts arose, more in the 1960s with the development of plate tectonics than with Wegener's early theory. Wegener gave us continental drift, but the 1960s gave us such important new concepts as seafloor spreading, transform fault, plate, subduction, lithosphere, and asthenosphere. The developments of the 1960s involved many more additions to geological vocabulary than did Wegener's theory.

Second, conceptual change is not merely a matter of adding new concepts, but also of *reorganizing* them. According to the first two theses on scientific revolutions stated in Chapter 1, revolutions involve transformations in the kind-hierarchies and part-hierarchies that are the main organizing backbone of conceptual systems. The geological revolution in the 1960s displays a striking transformation that involves both kind-relations and part-relations. During the 1950s, extensive surveys of the ocean floor began to cause problems for the old view that continents and seafloor are both parts of crust that differed primarily only in that the latter is submerged. This view was important for the contractionist alternative to Wegener's continental drift theory, which explained the similarity of flora and fauna in Africa and South America by postulating that there had been continental land bridges that subsided and became seafloor. Observations of the seafloor suggested, however, that seafloor and continents differed structurally and in composition, challenging the notion of land bridges: if the seafloor and the landmasses were fundamentally different, one would not expect that parts of the continents would occasionally sink and become seafloor.

Hess's (1962) proposals about seafloor spreading explained why the seafloor, which on his view was continuously being produced by molten material rising up at mid-ocean ridges, should be younger and different in constitution from continents. Once continents and seafloor were seen as more different than alike, it became more appropriate to think of seafloor and the continents as *kinds* rather than *parts* of crust. Moreover, the notion of crust became much less important, because what really mattered was that continental crust and oceanic crust were merely surface features of the huge plates that, according to plate tectonics, were the primary structural components of the earth's surface. In contrast to older views that saw the continents and seafloor as parts of a uniform crust, plate tectonics depicted continents and ocean floors as different kinds of surfaces of plates. *Crust* came to refer to the surfaces of plates, but played no dynamical role in geophysics. The geological revolution is the only case I know of a transformation *from* part-relations *to* kind-relations. This transformation is summarized in Figure 7.1, in which the top diagram shows the part-relations before plate tectonics, and the bottom diagram shows the part-relations and kind-relations after plate tectonics. In terms of the taxonomy of conceptual change in section 3.1, this transformation is best understood as an unusual case of tree switching, in which a fragment of the part-hierarchy becomes a fragment of the kind-hierarchy.

The conceptual transformation just described is only part of the general development that occurred in the twentieth century. To give a fuller picture of what happened, we shall now fill in some of the relevant history, presenting diagrammatic summaries of several important stages in the development of ideas: Wegener, his opponents, and the 1960s.

7.2.2 Wegener's Concepts

Figure 7.2 portrays the central concepts and explanatory relations of Wegener's system. Wegener wanted to account for the formation of mountains, the existence of fossils of land animals of the same species on landmasses separated by oceans, and the apparent fit of the continents on opposite sides of the Atlantic. He believed that the crust originally contained a protocontinent that was fractured by the two mechanisms he posited to account for continental

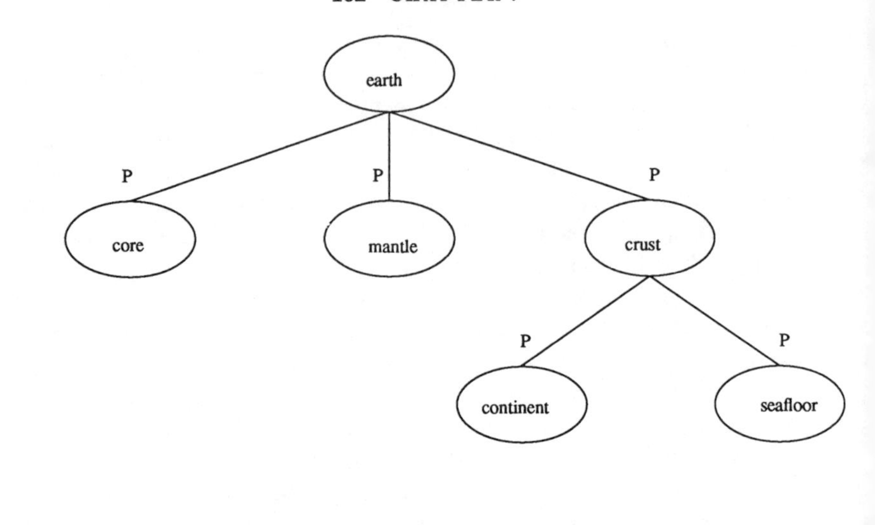

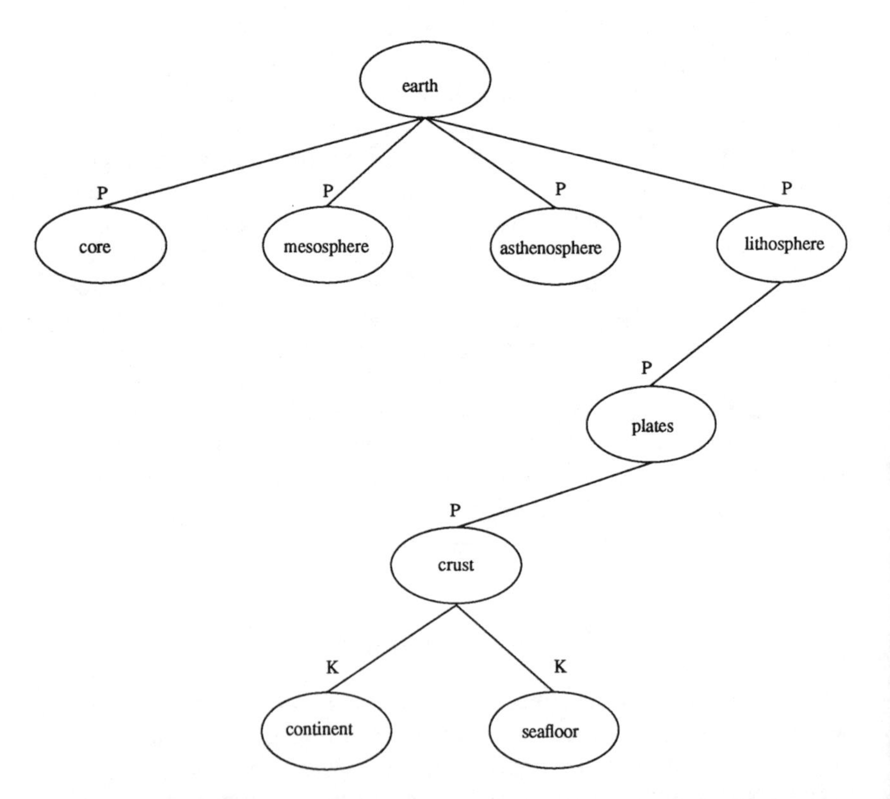

Figure 7.1. Part-relations and kind-relations before (top diagram) and after (bottom diagram) plate tectonics. Ellipses represent concepts. Part-relations are labeled P and kind-relations are labeled K, reading upwards. For example, in the top diagram, continent is a part of the earth's crust.

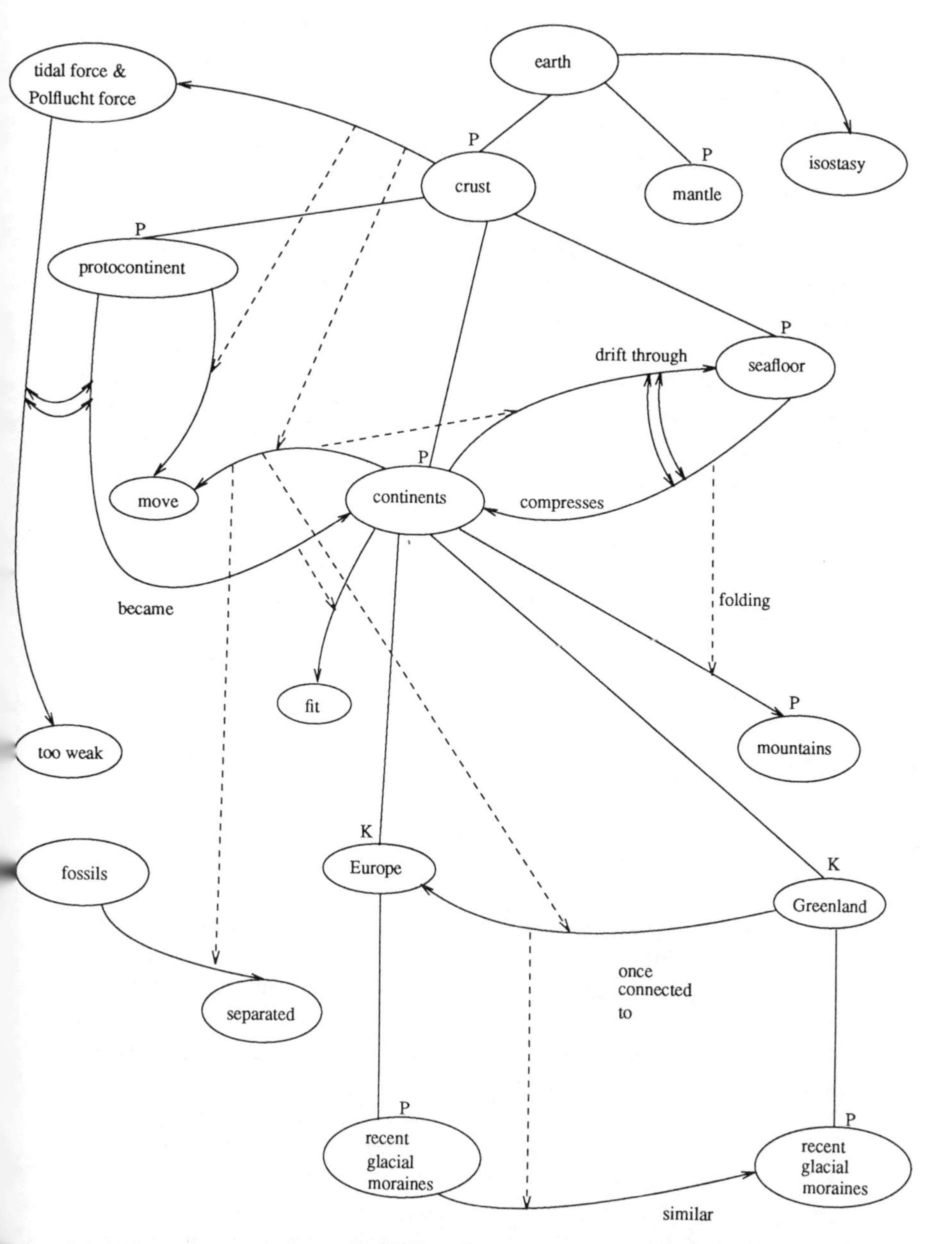

gure 7.2. Wegener's conceptual system. Part-relations are labeled P and kind-relations are ›eled K. Curved solid lines with arrows represent rules involving other kinds of relations. The ·tted lines indicate explanatory relations between propositions formed out of the concepts. For ample, that continents move explains why fossils are separated. Doubled lines with arrows on th ends join contradictory propositions.

motion: a tidal force that tended to drag continents westward; and the *Polflucht* force that moved continents away from the poles and toward the equator. Figure 7.2 and the other conceptual diagrams in this chapter are complicated and hard to follow, but I retain their complexity because they show the nature of conceptual systems more fully than the sketchy diagrams in other chapters. Even so, they only show a selection of the relevant concepts and their interconnections.

Figure 7.2 shows that Wegener's tidal and *Polflucht* forces were intended to be responsible for the motion of the continents and protocontinent, causing the protocontinent to split into the continents of today. Soon after Wegener's work was first published, it was pointed out that both of these forces were too weak by about five orders of magnitude to accomplish their intended effect. Wegener was able to explain the apparent "fit" of the continents, the creation of mountains, the separation of fossil species, and glacial phenomena; but the relationship between the seafloor and the continents was unclear. The continents had to be denser than the seafloor so they could move through it, but lighter than the seafloor so they could be compressed by it into mountains. Secondly, Wegener misjudged the relationship between certain glacial moraines found in both Europe and Greenland, assuming them to be related. Thus he conjectured an unreasonably recent date for the fission of Greenland and Europe, and based his calculations of the rate of drift on this date. Faulty longitudinal observations actually seemed to corroborate Wegener's claimed figure for the rate of drift, but later, more accurate observations in the 1930s showed no sign of drift more significant than statistical error.

7.2.3 Concepts of Wegener's Opponents

For Wegener's opponents, the basic force causing change in the earth's structure was the cooling of the earth, causing shrinking and lateral compression in the crust. Figure 7.3 provides a sketch of the conceptual system of Wegener's contractionist opponents (see van der Gracht et al. 1929). After buildup of stress, the crust fractured into blocks. This description is represented by the sequence of explanatory arrows at the upper left of Figure 7.3. One problem for a comprehensive account of the earth's geology was the presence in mountains of sedimentary rocks that had formed under water and contained aquatic fossils. The problem was solved by positing that sedimentary rocks formed over subsided blocks that were later elevated by further compression of the crust. This also explained the formation of the mountains themselves. Several pieces of evidence later claimed as support for continental drift found explanations within the contractionist system. For example, evidence for ancient glaciation in what are now temperate zones was explained by assuming that the earth's pole had wandered over the course of

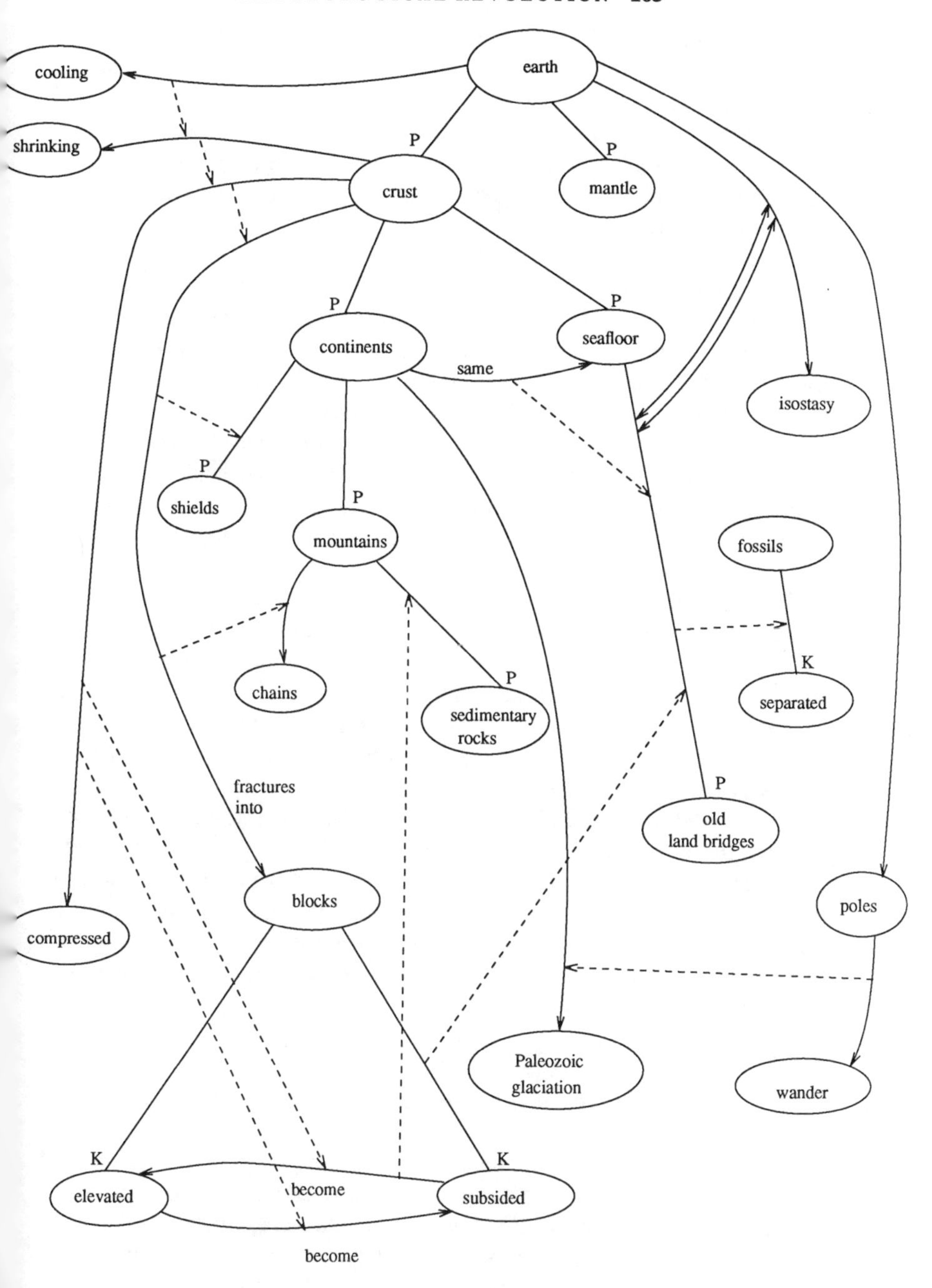

ure 7.3. Conceptual system of Wegener's critics. See Figure 7.2 for description of the notation.

time; the rotational axis of the earth had not always passed through the current North and South Poles.

The presence of nearly identical fossils of early land-based flora and fauna on opposite sides of oceans was explained by the prior existence of land bridges, land connections between continents that spanned the oceans. These land bridges are no longer in evidence since they were part of blocks that had subsided. The assumption that parts of the crust considered "land" could become parts of the crust considered "seafloor" presupposed that the continental crust and the seafloor were effectively identical in composition. Also, the ability of a large mass of crust to significantly change its relative altitude was in conflict with an accepted principle of geophysics. The principle of isostasy asserts that the crust is usually in gravitational equilibrium. A large piece of the crust that has not recently been disturbed has equal forces holding it up and pulling it down: buoyancy with respect to denser strata below, and gravity. The conflict between the hypotheses of isostasy and land bridges was tolerated because they belonged to different disciplines. Only geophysicists were concerned with the validity of isostasy, and only geologists were concerned with the existence of land bridges.

The contractionist program was not stagnant after the initial rout of Wegener. By 1959, it had been expanded to account for most of the major features of the earth's surface. The authors of a major contractionist text (Jacobs, Russell, and Wilson 1959) continued the geophysical tradition of ignoring the issue of land bridges in their discussion of isostasy, but were aware of one problem in their synthesis: the average heatflow from the ocean floor and the continents was the same, even though the greater amount of radioactive material in the continents should produce higher continental heatflow. This conflict was discussed but not resolved by the text. Contemporary accounts that did discuss the issue assumed that isostasy was a rough approximation and that land bridges could exist for significant amounts of time, or that the forces governing continental uplift and subsidence were stronger than isostasy, or that the land bridges were little more than isthmuses not large enough to be subject to isostasy.

7.2.4 Concepts of Hess and Plate Tectonics

The 1950s saw a massive accumulation of data, especially about the seafloor. The ocean floor's central ridges, known since the laying of the transoceanic cables, were found to have rifts down the middle. These rifts were almost invariably locations of high earthquake activity, and maps of earthquake epicenters were used to detect mid-ocean rifts. Heatflow profiles, measured in terms of units of energy per unit of area, became available. The seafloor heatflow profiles peaked at the central rifts, but there was no theoretical attempt at explaining the peaks.

Magnetic data also became significant in the 1950s. When some molten rocks cool, they are magnetized by the earth's magnetic field. Like a bar magnet, they have a direction associated with them, pointing toward the location of the north magnetic pole at the time the rocks were magnetized. The data suggested that the earth's magnetic field had periodically reversed over time. Although rocks magnetized in the opposite direction from the current magnetic field of the earth had been known since the 1920s, the reality of periodic reversals was not accepted until the 1950s.

In 1959, Harry Hess first proposed his theory of seafloor spreading, which was eventually published in 1962. Robert Dietz, who coined the term "seafloor spreading," advanced similar ideas but gave Hess priority. Hess's theory took account of the mass of new data gathered about the seafloor during the 1950s, and attempted to explain it in a way that supported the hypothesis of continental drift. Figure 7.4 gives a rough picture of the conceptual relations in Hess's theory.

Hess hypothesized that convection in the mantle caused rising areas of mantle material to surface at the crust at mid-ocean ridges. This rising material spread open the existing seafloor, cooled, and became new seafloor. In turn, old seafloor was consumed by the mantle at subduction zones located at oceanic trenches. One major difference between Hess's work and previous views was that the ocean floor was no longer considered part of the crust; since it was composed of cooling mantle, the seafloor was seen as "crustless" and the only crustal material was continental. Hence the diagram has no part link joining *crust* and *seafloor*. Note the central importance of the relation between the concepts *seafloor* and *spreads* in the diagram; the proposition expressing this relation was used to explain most of the major features of the seafloor and continents.

Shortly after Hess's paper appeared, new magnetic evidence seemed to confirm it. Strata on land revealed that the earth's magnetic field had reversed polarity in the past. Magnetic profiles of the seafloor revealed linear patterns parallel to the mid-ocean ridges, and cores of sediment taken from the seafloor also gave evidence of magnetic reversals. Vine and Matthews (1963) hypothesized that if new seafloor was being created at the mid-ocean ridges, then there should be patterns of magnetic stripes parallel to the ridges, which record the direction of the earth's magnetic field at the time the seafloor was being created. After many ambiguous measurements, relatively clear data were obtained to confirm the Vine-Matthews hypothesis that seafloor was constantly forming and recording the direction of the earth's magnetic field as it cooled. The time-scales of magnetic reversals obtained by three different methods—land-based sampling, horizontal striping around the mid-ocean ridges, and reversals recorded in columns of sediment above the seafloor—were in general agreement. Other concerns such as mountain building were also addressed. Mountains were said to be created when continents crashed into each other or resisted the motion of the seafloor. All the paleontological

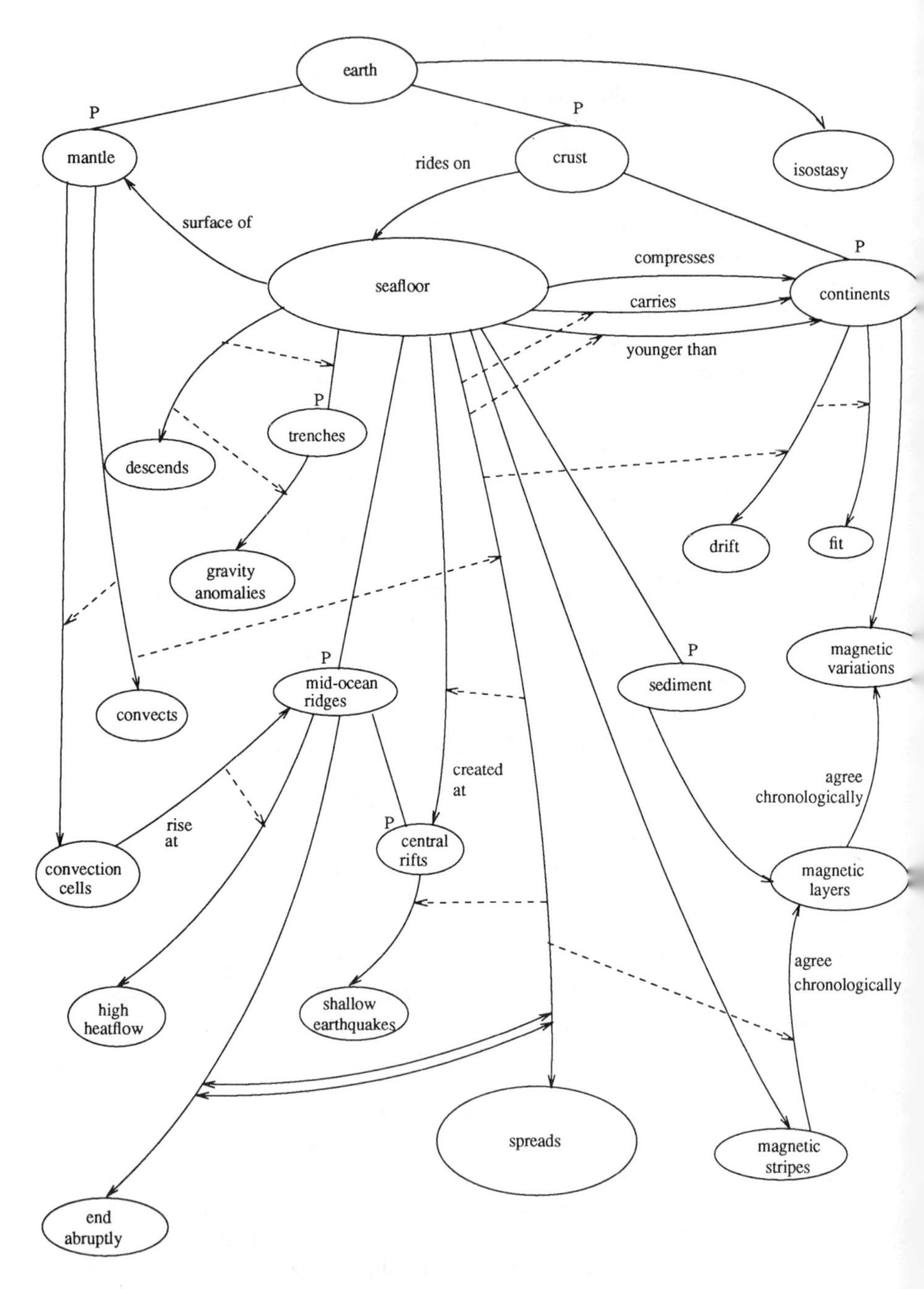

Figure 7.4. Conceptual system of Hess. See Figure 7.2 for description of the notation.

and geological data that originally were explained by continental drift also supported seafloor spreading since seafloor spreading allowed for drift.

Many of the facts explained by seafloor spreading were relatively new pieces of evidence discovered in the 1950s: the high heatflow of the mid-ocean rift, and gravitational and magnetic anomalies. Hess's main goal was to account for features of the seafloor, not to defend continental drift, but his theory implied that continental drift had occurred. One minor internal conflict appeared for Hess's theory: some mid-ocean ridges ended abruptly in relatively undisturbed seafloor. If seafloor was being created by spreading at mid-ocean ridges, how could these ridges come to an end?

Hess's hypothesis of seafloor spreading showed the benefits of a theory that embraced continental drift; the task remained to elaborate it beyond the level of the "geopoetry" proposed by Hess. In 1965, J. Tuzo Wilson proposed the existence of a new kind of fault that he called the *transform fault*. Study of Hess's work had led him to realize that if Hess was correct, a hitherto unobserved type of fault would be required to allow motion to parts of the earth's crust. Along with the existence of transform faults, Wilson described the crust as being divided into large *plates* that were composed of both seafloor and continental material; the mineralogical differences of the two types of crustal material were now considered less important than their structural unity. Plates could be bounded only by mid-ocean ridges, trenches or subduction zones, and transform faults.

Geophysicists borrowed a new organization of the interior of the earth from seismologists. Seismic waves exhibited a change in behavior at a depth of about 100 kilometers. Immediately below this level was a relatively plastic zone over which plates could slip with less resistance than would otherwise be suspected. Thus the depth of 100 kilometers could be interpreted as the thickness of the plates. The plates were the structural components of the lithosphere, which was the part of the earth's surface that moved; the asthenosphere was the portion of the mantle in which convection occurred. Once again the crust included seafloor; it was the surface of the lithosphere and was thus broken into plates as well. Although many of the ideas of Hess were preserved in plate tectonics, their organization was different, and different points were emphasized. The development of plate tectonics made the continents gradually less important for geophysics than the seafloor, and both concepts became less significant than the concept of plates. The mature theory of plate tectonics arose when Jason Morgan (1968) developed a mathematical framework for plate tectonics that yielded quantitative predictions. Figure 7.5 depicts the qualitative concepts of plate tectonics.

The richness of the history just sketched shows that much work remains to be done in modeling the conceptual development of the key ideas in plate tectonics. Instead of pursuing that project further here, however, I want to move on and address questions about the *evaluation* of geological theories.

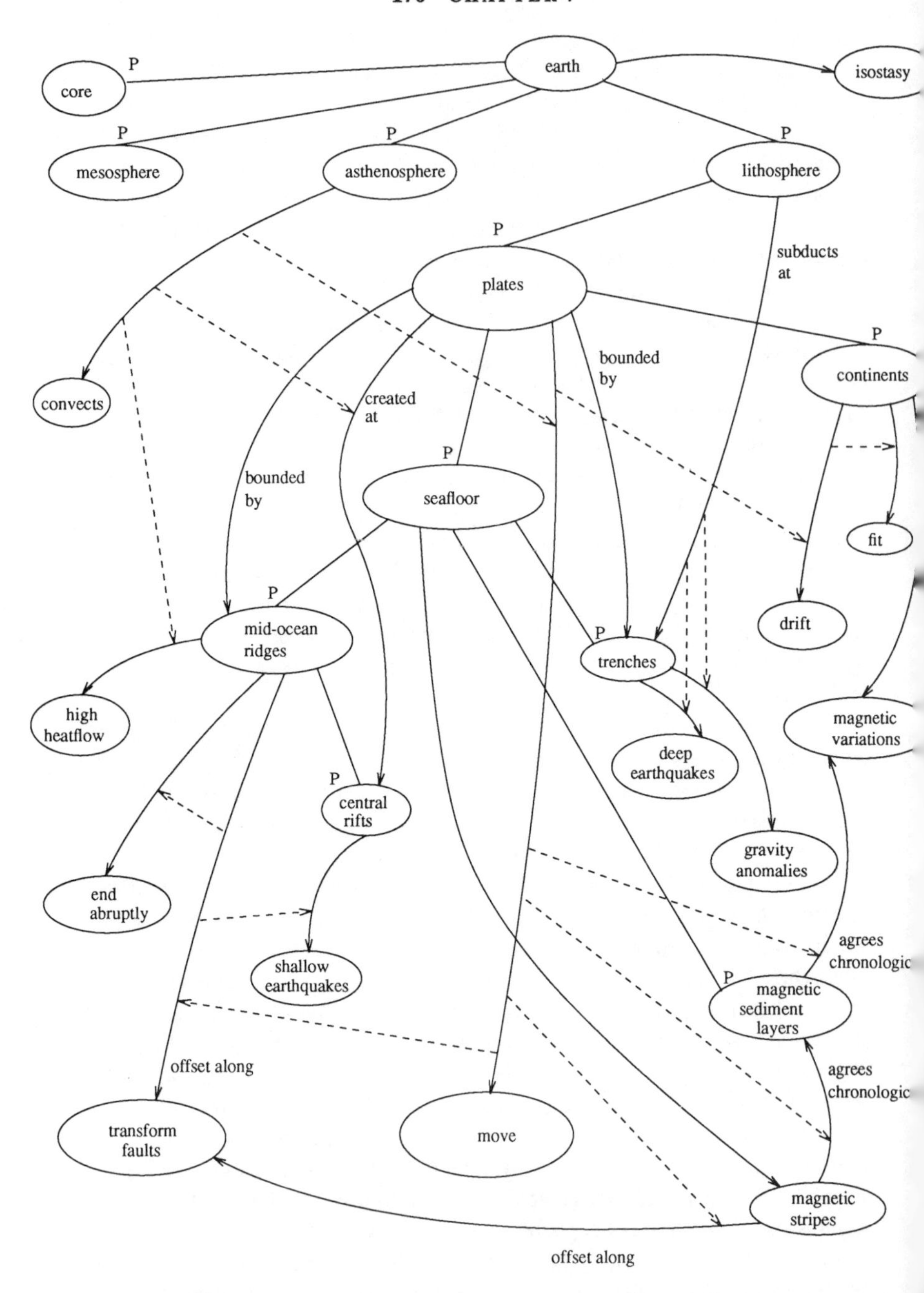

Figure 7.5. Conceptual system of plate tectonics. See Figure 7.2 for description of the notation

7.3 EVALUATING THEORIES OF CONTINENTAL DRIFT AND PLATE TECTONICS

So far, we have been addressing questions of discovery and conceptual development. But computational modeling of scientific thinking should also provide insights into how theories are evaluated and accepted. The key historical question that needs to be answered is:

> *Why was Wegener's theory of continental drift largely rejected in the 1920s, and why, in contrast, were the new ideas about seafloor spreading and plate tectonics largely accepted in the 1960s?*

I shall answer these questions using the theory of explanatory coherence presented in Chapter 4.

7.3.1 The Coherence of Continental Drift

Between 1910, when he first formed the hypothesis, and 1929, when the fourth German edition of his book was published, Wegener amassed a wide variety of kinds of evidence explicable on the assumption that the continents were once joined and have drifted apart. He provided a unified explanation of many geological and paleontological phenomena, and he attempted to explain how continental drift could take place through the mechanisms described in section 7.2.2. Thus the hypothesis of continental drift gains coherence both from what it explains and its being explained.

Wegener's argument for continental drift was explicitly comparative in his attack on contrary contractionist views that had been used to explain related phenomena. According to the widely accepted contraction theory, a cooling, shrinking earth had a crust that buckled, fractured into blocks, and folded as the earth shrank. The interactions of these blocks during cooling were supposed to explain most of the significant geological and paleontological features of the earth. Blocks that overrode others, were raised, or tipped, accounted for mountain formation. Blocks that subsided and were covered by ocean explained the absence of visible land bridges that paleontologists had postulated to account for the widespread dispersal of many land-based fossil animal and plant species over several of today's continents. Wegener saw the contractionist hypothesis as contradicting the geophysical theory of isostasy, which held that the continents were in gravitational equilibrium and thus could not rise or sink significantly.

Analysis of the argument of Wegener (1966) produced forty propositions that constituted for him the principal evidence and explanatory hypotheses. Tables 7.1 and 7.2 show the input given to ECHO. (Because of their length, all tables for this chapter are placed in an appendix, section 7.5. Propositions labeled "NE" are pieces of "negative evidence" that contradict evidence.) ECHO takes these inputs and creates the network of excitatory and inhibitory links shown in Figure 7.6. Notice the large number of explanations participated in by hypotheses W5 (the continents were once connected) and W11 (continents are moving). In turn, these two propositions are explained by the mechanisms postulated in W8 and W9. ECHO notes the following competitors:

C8 competes with C7 because of (E8).
W10 competes with C7 because of (E8).
W10 competes with C10 because of (E8).
W11 competes with C1 because of (E15).
W11 competes with C9 because of (E15).
W11 competes with C6 because of (E5).
W11 competes with C4 because of (E2 E3).
W11 competes with C5 because of (E2 E3 E5).
W11 competes with C3 because of (E2 E3).
W2 competes with C9 because of (E18).
W4 competes with C10 because of (E12).
W4 competes with C3 because of (E2 E3).
W4 competes with C5 because of (E2 E3).
W4 competes with C4 because of (E2 E3).
W5 competes with C6 because of (E5).
W5 competes with C5 because of (E5).
W6 competes with C9 because of (E15).
W6 competes with C1 because of (E15).
W8 competes with C4 because of (E2).
W8 competes with C5 because of (E2).
W8 competes with C3 because of (E2).
W9 competes with C3 because of (E3).
W9 competes with C5 because of (E3).
W9 competes with C4 because of (E3).
W9 competes with C1 because of (E15).
W9 competes with C9 because of (E15).

Figure 7.7 shows the connectivity of a sample unit.

When the network is run, repeated updates of the activations of the units leads to the acceptance of the Wegener hypotheses and the rejection of most of the contractionist hypotheses. Primarily because Wegener's hypotheses have more excitatory links with the evidence, their units gradually win out over the opposing theory of contraction, whose units lose activation. Figure 7.8 shows the activation histories of all units representing hypotheses. Some of the contractionist units, C8, C9, and C10, finish with high activation, since

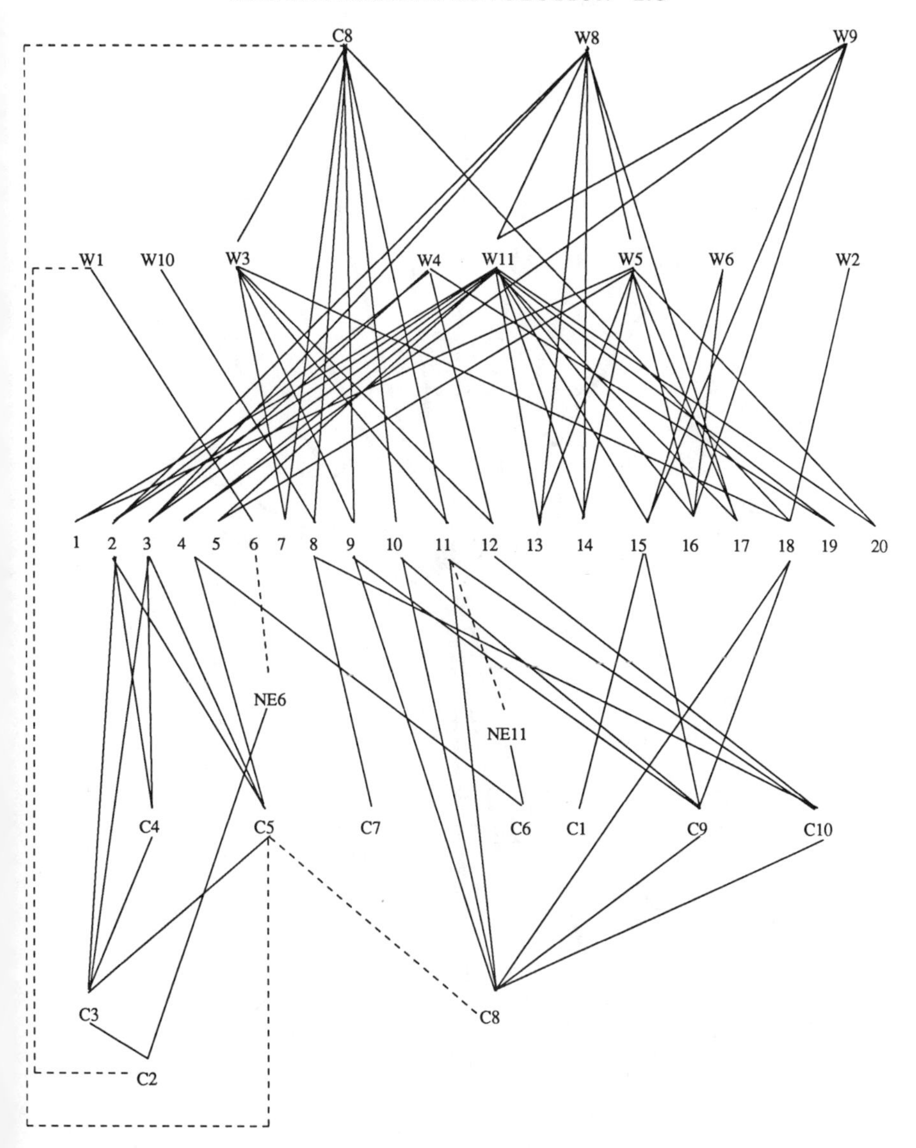

Figure 7.6. The explanatory coherence of continental drift, from Wegener's point of view. 1–20 represent evidence units E1–E20. NE6 and NE11 are pieces of negative evidence contradicting evidence statements. W1–W11 are Wegener hypotheses, and C1–C10 are contractionist hypotheses, except that C8, isostasy, is common to both sides and is shown twice. Solid lines indicate excitatory links representing coherence, while dotted lines indicate inhibitory links established by contradictions. Inhibitory links established through competition are not shown. Also not shown are the special evidence unit and the links between it and the evidence unit.

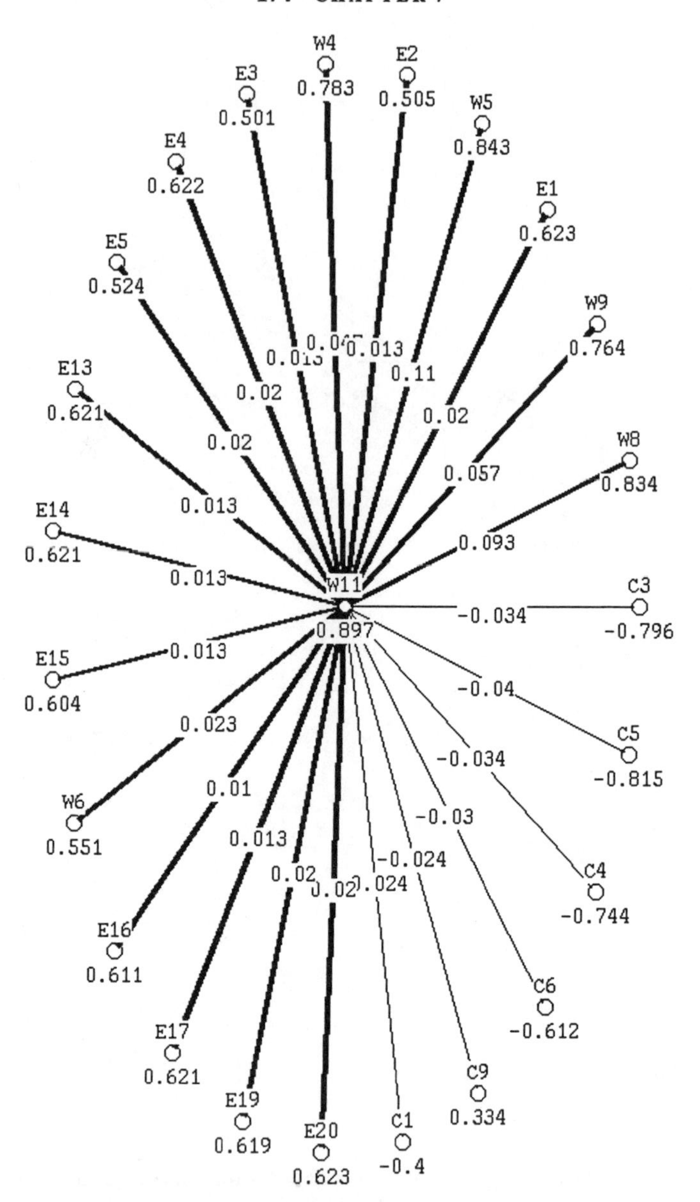

Figure 7.7. Connectivity of unit W11. The numbers under the units are their activation values after the unit has settled. Thick lines indicate excitatory links, while thin lines indicate inhibitory links. Numbers on the lines indicate the weights on the links.

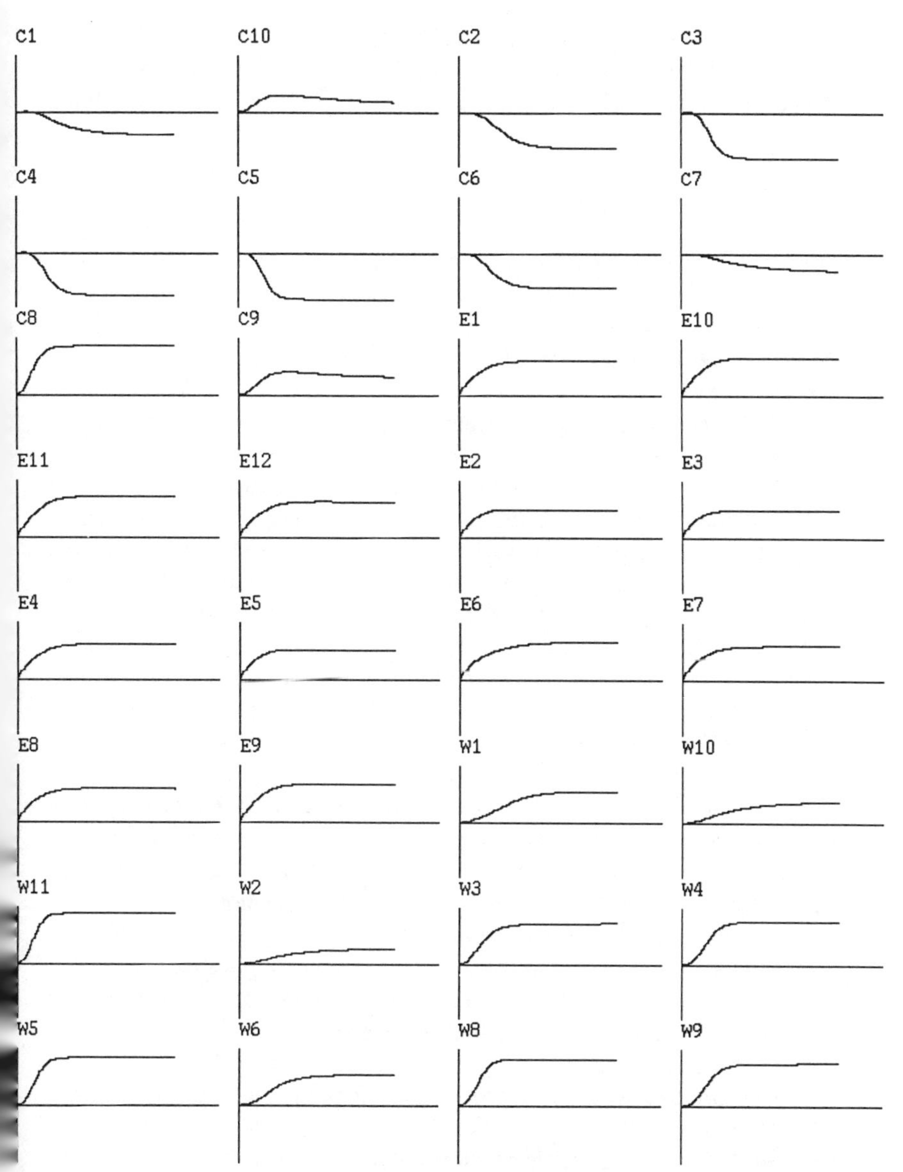

Figure 7.8. Activation histories of propositions in Wegener simulation. Each graph shows the activation of a unit over 99 cycles of updating, on a scale of –1 to 1, with the horizontal line indicating the initial activation of 0. Note that drift hypotheses W1–W11 are all accepted, while most of the contractionist hypotheses are rejected.

they are not in conflict with the views of Wegener, who agreed with the theory of isostasy. Let me stress that this simulation models Wegener's own belief system as put forward in his book; his later critics require separate treatment.

7.3.2 The Rejection of Continental Drift

From Wegener's own perspective, the theory of continental drift is highly coherent, explaining a broad range of facts. But most geologists did not share Wegener's conclusion that continental drift should be accepted. In 1928, the American Association of Petroleum Geologists published a symposium on Wegener's views (van der Gracht et al. 1929). Most (but not all) symposiasts were highly critical of Wegener. We have accordingly compiled a composite picture of Wegener's opponents by amalgamating the various criticisms of his account, most of which asserted that his hypotheses contradicted well established evidence. The input to ECHO based on our analysis of the criticisms Wegener received is presented (section 7.5) in Tables 7.3 and 7.4, which show the relevant propositions, explanatory relations, contradiction relations, and evidence. Given this input ECHO produces the network shown in Figure 7.9. Note the large number of NE statements, pieces of "negative evidence," which are consequences of Wegener's view that, according to his critics, contradict actual evidence. Hypotheses W1–W6 are hypotheses of Wegener, while C1–C7 are hypotheses of the alternative contractionist view. CW1 is common to both Wegener and the contractionists.

The result of running this network, updating the activation of each node based on the activation of the nodes with which it is connected, is that the contractionist hypotheses are accepted and the Wegener hypotheses that contradict them are rejected. Figure 7.10 shows the activation histories of the units representing hypotheses. Wegener's critics rejected his views for several reasons. First, they saw his hypotheses as having implications contradictory to much of the available evidence; hence the raft of NE statements in Table 7.4. Second, they were satisfied with the available contractionist explanations of the evidence.

Ronald Giere (1988) has offered a very different account of why Wegener favored his theory and why his critics were not convinced. Giere contends that scientific theory choice should be understood in terms of a "satisficing" model of decision making (Simon 1945). According to Simon, decision makers do not optimize; instead, they "satisfice" by choosing some less than optimal decision that is nevertheless satisfactory according to some relaxed criteria. Giere says that the choice faced by Wegener and his opponents was whether to accept mobilism (continental drift) and reject the prevailing stabilist views, or to reject stabilism and accept mobilism. The decision is based on both

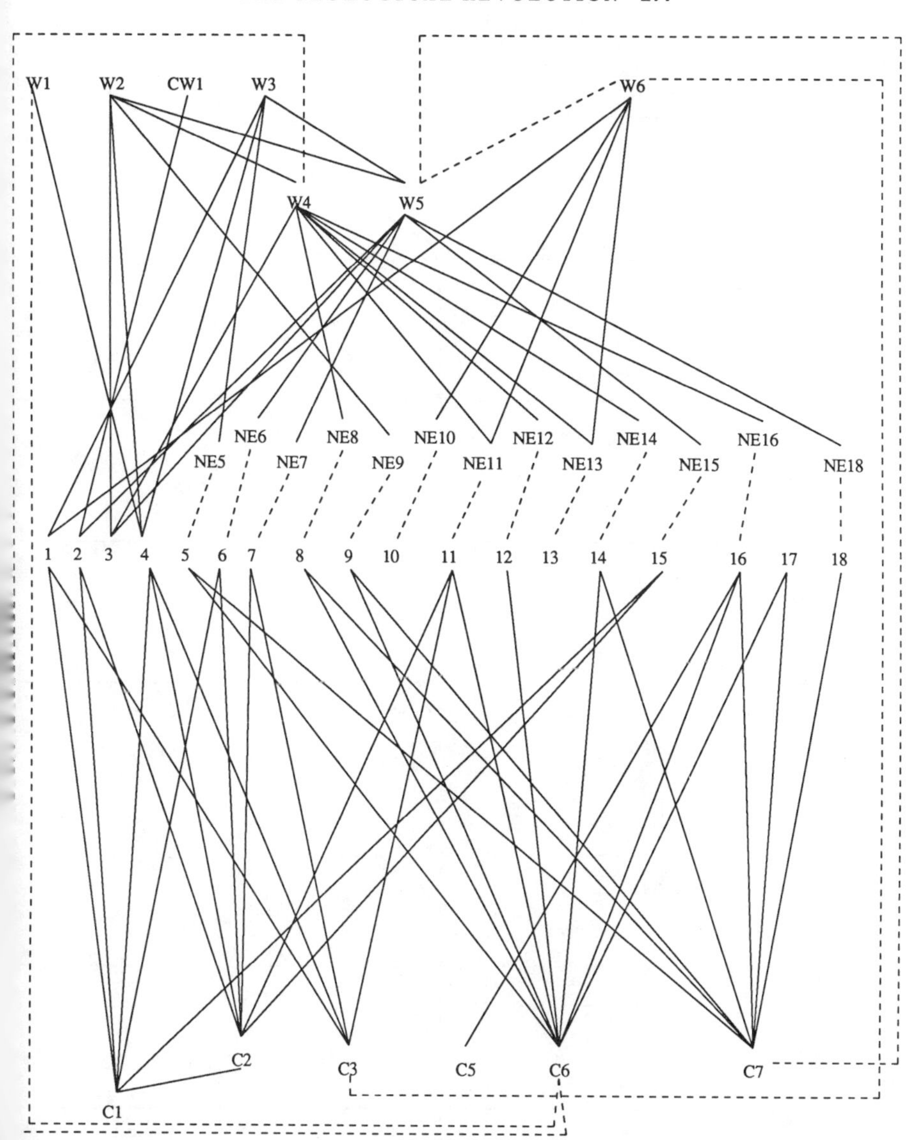

Figure 7.9. The explanatory coherence of contractionist hypotheses versus Wegener's hypotheses, from the point of view of Wegener's critics. W1–W6 represent Wegener's hypotheses, while C1–C7 represent contractionist hypotheses. CW1 is common to both. Numbers 1–18 represent the pieces of evidence labeled E1–E18 in Table 7.3. NE5–NE18 are pieces of negative evidence that contradict pieces of evidence. Solid lines indicate excitatory links representing coherence, while dotted lines indicate inhibitory links representing incoherence. Inhibitory links deriving from competition are not shown.

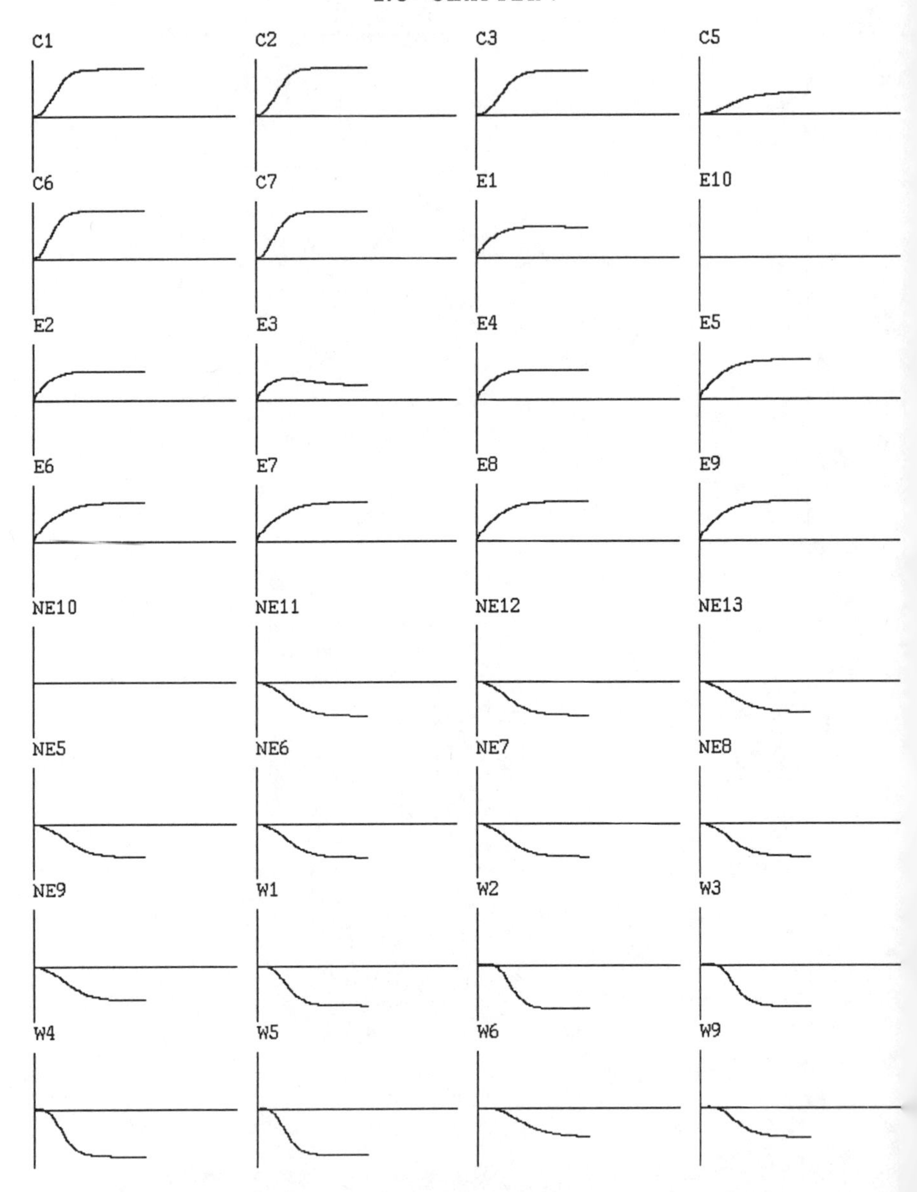

Figure 7.10. Activation histories of selected units in simulation of Wegener's opponents. Each graph shows the activation of a unit over 70 cycles of updating, on a scale of –1 to 1, with the horizontal line indicating the initial activation of 0. Note that contractionist hypotheses C1–C7 are all accepted, while drift hypotheses W1–W6 are all rejected.

epistemic and nonepistemic interests of the participants. The epistemic ones concern conditional probabilities; for example, if the existence of a good fit between the shapes of the continents on either side of the Atlantic is more probable given continental drift, then that counts in its favor. The nonepistemic interests include the personal concerns of the participants. For example, Giere says that the most adamant critics of Wegener were those who were most entrenched in the professional community of geologists, while Wegener was an outsider and had little to lose in professional standing by taking an audacious position.

The explanatory coherence account of Wegener and his opponents is superior to Giere's in several ways. First, his talk of probability is metaphorical at best, since he gives no account of how one could go about assessing conditional probabilities involving qualitative hypotheses such as continental drift. In contrast, Wegener and his opponents do explicitly speak of what can and cannot be explained using continental drift. Second, there is no reason to invoke nonepistemic interests in accounting for the difference in conclusions between Wegener and his opponents. They were all assessing the hypotheses with respect to coherence with their overall beliefs. Third, Giere's model provides no way of integrating the factors that he claims are operating in this case. In contrast, ECHO shows how disparate claims involving numerous explanatory relations can be collectively assessed using a connectionist algorithm for maximizing explanatory coherence. I therefore conclude that early disputes about continental drift are better understood in terms of explanatory coherence than in the decision-theoretic terms favored by Giere.

7.3.3 The Acceptance of Plate Tectonics

Continental drift fared much better several decades later. ECHO has been used in an analysis of the seminal paper of Harry Hess (1962) that first developed the concept of seafloor spreading. Hess explicitly presented his views as a set of propositions and indicated numerous phenomena that seafloor spreading can explain. His proposition that the seafloor spreads (S16) directly competes with the contractionist proposition that the crust is contracting (C2) since they function as opposing accounts of many crustal features. Hess himself did not do a systematic comparison of his account against the contractionist theory; because that theory was undoubtedly in the background, the ECHO analysis includes standard 1950s contractionist explanations, taken mostly from Wilson (1954). The input given to ECHO is shown in Tables 7.5 and 7.6 (see the appendix). Figure 7.11 shows the coherence relations of the most important hypotheses of Hess and the contractionists. C2, asserting that the earth contracts, contradicts S4, the main hypothesis of the seafloor spreading

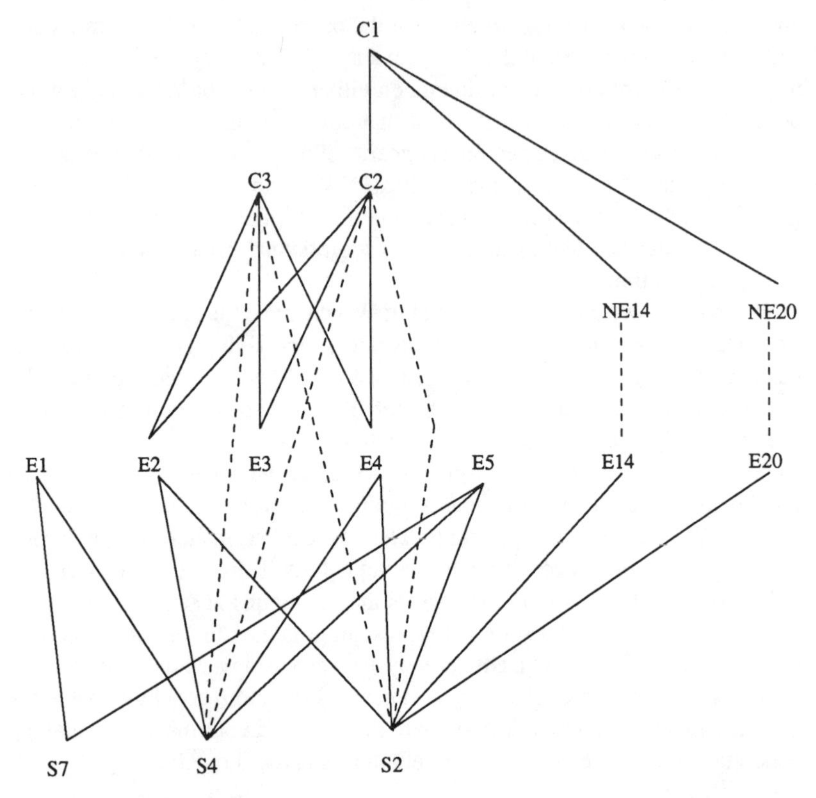

Figure 7.11. Partial depiction of the explanatory coherence of Hess's theory of seafloor spreading. The notation is the same as in Figure 7.9. Shown are inhibitory links deriving from both contradiction (C2 and S4) and competition (C2 and S2, C3 and S2, C3 and S4).

model. Several hypotheses of the contractionists are explained by C1, the hypothesis that the earth's mantle has been cooling. But this hypothesis implies "negative evidence" NE14 and NE20 about the measured heatflow of the crust, which contradict actual observations represented by E14 and E20. The spread model explains these observations. The contractionist and spread models have about the same number of mutually coherent hypotheses. As the abbreviated diagram of Figure 7.11 suggests, the spread model derives its advantage by explaining pieces of evidence that the contractionist model does not, and by explaining actual observations that contradict the implications of the contractionist model. Of the propositions shown in Figure 7.11, ECHO finds the following hypotheses to be competitive:

C2 competes with S4 because of (E2 E4 E9).
C2 competes with S2 because of (E2 E3 E4 E9 E12 E23).

C3 competes with S2 because of (E2 E3 E4 E9 E12 E23).
C3 competes with S4 because of (E2 E4 E9).

When the network produced by the input in Tables 7.5 and 7.6 settles, Hess's seafloor spreading hypotheses win out, while those contractionist hypotheses that contradict Hess are deactivated. Proposition C8, that the earth's magnetic pole has wandered over time, is not rejected, but it is consistent with Hess's position. From Hess's perspective, seafloor spreading and its consequences for continental drift should be accepted as possessing more explanatory coherence than the contractionist theory.

This ECHO run only models Hess, not the entire geological community, which did not immediately accept Hess's views. For some geologists not as familiar with seafloor phenomena as Hess, later developments were important in making seafloor spreading acceptable. I have already mentioned developments such as the confirmation of the Vine-Matthews hypothesis by the *Eltanin* magnetic data and Wilson's ideas of transform faults and plates. In my view, the importance of the magnetic data has been overemphasized by philosophers of science who think that prediction and confirmation is more important than explanatory coherence (Frankel 1979; Giere 1988). For some geologists such as Menard (1986), the magnetic data were important in tipping the balance in favor of the seafloor spreading hypothesis. But for other leaders in the field such as Hess (1962), Dietz (1961), and Wilson (1963), seafloor spreading was already accepted before the magnetic data was in. Wilson is a particularly interesting example, since he had been a leading exponent of the contractionist view, but was an early convert to the new ideas. The simulation of Hess can be expanded, adding explanation of the magnetic data by seafloor spreading. The additional excitation increases the activation of the node representing seafloor spreading in comparison to that possessed by its competitors, but the improvement is slight. The *Eltanin* data by no means constituted a crucial experiment.

We can now answer the question with which this section began: Why was Wegener's theory of continental drift rejected while the 1960s theory was accepted? The short answer is that the 1960s theory had much more explanatory coherence than Wegener's theory. There are three principal reasons for the relative superiority of the later theory. First, by the 1960s much more evidence had been collected that was explained by hypotheses of seafloor spreading and continental drift. Many facts had been brought to light by the explorations of the seafloor in the 1950s that were much more easily explained by seafloor spreading than by the contractionist theory, even though the latter had seemed superior to Wegener's theory. Second, Hess's theory of seafloor spreading, as well as the fully developed theory of plate tectonics, did not have the incoherence that Wegener's theory had with various geological facts and physical theory. The mechanisms postulated for seafloor spreading and plate movement do not contradict accepted principles in the way that Wegener's views of the continents plowing through the seafloor did. The third reason for the relative preferability of the 1960s theory over Wegener's is that the alternative contractionist theory was on the wane. Accumulating geophysical evidence, such as detailed information about the amount of heat-generating radioactivity in the earth's crust, made the view that the crust was cooling less tenable. Thus explanatory coherence theory and ECHO appear adequate for accounting for the evaluation and acceptance of theories in the geological revolution.

What is the relation between plate tectonics and the contractionist theory that it succeeded? As with the theories of Lavoisier and Darwin, plate tectonics absorbed too little of contractionist theory to be described as incorporating or sublating it. Instead, we should say once again that plate tectonics *supplanted* the previous thcory, since it involved rejection of central hypotheses of the contractionist theory. Moreover, as we saw in section 7.2, the adoption of plate tectonics required substantial reorganization in the conceptual hierarchies involving kind-relations and part-relations. There appears, however, to have been considerable cumulation of evidence and methods.

Finally, what strands of explanation were used by geologists arguing about continental drift and seafloor spreading? Until Morgan's mathematical theory of plate tectonics, geological explanations were at best loosely deductive. They appear to have a schematic aspect, since similar patterns of explanation, involving continental drift or seafloor spreading, were repeatedly applied. Common to all the geological explanations was an attempt to provide a causal mechanism sufficient to produce the observed phenomena, so explanations in the geological revolution are probably best characterized as causal.

7.4 SUMMARY

Cognitive mechanisms of abduction and conceptual combination contributed to the discovery of the theory of continental drift, which was the forerunner of the theory of plate tectonics. The acceptance of plate tectonics depended on new concepts such as seafloor spreading. Acceptance also required a conceptualization of continents and ocean floors as different kinds of surface and parts of plates, instead of as parts of the earth's crust. The theory of explanatory coherence and ECHO can be used to model Wegener's case for continental drift and its criticism by the dominant opposition. Explanatory coherence theory makes it clear why Hess's theory of seafloor spreading and the theory of plate tectonics that grew out of it were much more acceptable than Wegener's views. The later theories explained much new evidence and avoided numerous incoherent aspects of Wegener's theory.

7.5 APPENDIX

Table 7.1
Propositions for Simulation of Wegener

Evidence:

E1	The shape of the Atlantic coastlines match.
E2	There are several North-South mountain chains.
E3	Major folding also occurs along East-West lines.
E4	Tapering North-South blocks curve to the East.
E5	Many species of fossils are represented on both sides of the Atlantic.
E6	The current slow heat loss of the earth requires relatively low temperatures for the ancient crust.
NE6	The earth's crust was significantly hotter in the ancient past.
E7	The ocean floor is largely level, and continental surfaces are generally 5 km. above it.
E8	Shallow-sea but no deep-sea fossils are found on land.
E9	Scandinavia appears to be rising with respect to the seafloor.
E10	No anomalous gravity measurements have been observed near mountains.
E11	The seafloor and continents are of different composition.
NE11	Large amounts of continental crust can be found under the oceans.
E12	There are no folded mountains on the seafloor.
E13	Strata on opposite sides of the Atlantic match.
E14	There are no ancient species in the Atlantic, but there are some in the Pacific.
E15	During the Cretaceous period, Spitzbergen was tropical and Central Africa (90 deg. south) was polar.
E16	Seemingly contradictory evidence about location of earlier poles.
E17	Measurement of the position of Greenland indicates motion in excess of standard error.
E18	Gravity deficiencies have been measured at oceanic trenches.
E19	Most ancient rocks are greatly folded.
E20	The Atlantic is earthquake-free, but the Pacific has many.

Contractionist hypotheses:

C1	The geographic poles have moved with respect to the crust.
C2	The earth has been contracting since birth due to cooling.
C3	The crust is compressed and fractures into blocks.
C4	Blocks can override one another.
C5	Blocks can subside and become flooded or rise and tip.
C6	Land bridges once connected the continents.

C7	Isthmus connections served as land bridges.
C8	Large sections of the crust are in gravitational equilibrium (Isostasy).
C9	The crust swims in a layer of magma.
C10	The continents do not move vertically on a large scale.

Wegener hypotheses:

W1	The heat content of the earth is rising.
W2	The seafloor is depressed around continents by their weight.
W3	Crustal blocks swim in magma with only top 5 km. showing.
W4	The resistance of cool oceanic crust to continental movement causes mountain formation.
W5	The continents were once all connected.
W6	The earth's pole of rotation has moved with respect to the crust.
W8	Tidal forces pull apart the continents and move them westward.
W9	Polfluchtkraft moves continents toward the equator.
W10	Shallow seas may once have been isthmus connections.
W11	Continents move with respect to each other.

Table 7.2

Explanations and Contradictions for Simulation of Wegener

Explanations:

(explain (C6) NE11)
(explain (C8) W3)
(explain (C2) NE6)
(explain (C2) C3)
(explain (C3) C5)
(explain (C3) C4)
(explain (C8) C9)
(explain (C8) C10)
(explain (W8 W9) W11)
(explain (W8) W5)
(explain (W11 W5) E1)
(explain (W11 W4 W8) E2)
(explain (W11 W4 W9) E3)
(explain (W11 W8) E4)
(explain (W11 W5) E5)
(explain (W1) E6)
(explain (W3 C8) E7)
(explain (C8 W10) E8)
(explain (W3 C8) E9)
(explain (C8) E10)
(explain (W3 C8) E11)
(explain (W3 W4) E12)
(explain (W11 W5 W8) E13)
(explain (W11 W5 W8) E14)
(explain (W11 W6 W9) E15)
(explain (W11 W5 W6 W9) E16)
(explain (W11 W5 W8) E17)
(explain (W3 C8 W2) E18)
(explain (W11 W4) E19)
(explain (W11 W5) E20)
(explain (C3 C5 C4) E2)
(explain (C3 C5 C4) E3)
(explain (C6 C5) E5)
(explain (C10 C7) E8)
(explain (C8 C9) E9)
(explain (C8 C9) E10)
(explain (C10 C8) E11)
(explain (C10) E12)
(explain (C1 C9) E15)
(explain (C8 C9) E18)

Contradictions:

(contradict E6 NE6)
(contradict E11 NE11)
(contradict C5 C8)
(contradict C2 W1)

Data:

(data (E1 E2 E3 E4 E5 E6 E7 E8 E9 E10 E11 E12 E13 E14 E15 E16 E17 E18 E19 E20))

Table 7.3
Propositions for Simulation of Wegener's Opponents

Evidence:

E1	Mountains occur in chains, especially in the Tertiary period.
E2	Large shields on the continents are flat and stable.
E3	The outlines of continents on the opposite sides of the Atlantic are similar.
E4	Many mountains were formed prior to the Cretaceous period.
E5	North America and Eurasia are still north of the equator.
NE5	Continents that were originally north of the equator should be equatorial by now.
E6	Many earthquakes occur under the oceans.
NE6	A plastic ocean floor would not exhibit signs of stress such as earthquakes.
E7	There are trenches and ridges on the ocean bottom.
NE7	A plastic ocean floor would gradually lose its surface features.
E8	The continents bordering the Atlantic can't be made to fit together consistently.
NE8	The parts of an early supercontinent would fit together very well.
E9	There is no faulting due to tensile strain on the eastern edges of continents.
NE9	The drag on westward-moving continents would create stress faults in their eastern edges.
E10	There are no mountains on the ocean floor.
NE10	Rigid ocean floors would also be compressed and form mountains.
E11	Pre-Cretaceous mountain chains parallel current coastlines.
NE11	Features associated with the Atlantic coastlines would date from the Cretaceous or later periods.
E12	Epi-continental seas are ancient.
NE12	Features associated with continental boundaries would date from the Cretaceous or later periods.
E13	Australia experienced folding before but not after the Cretaceous period.

NE13 Australia would exhibit Cretaceous and later folding and mountain building.
E14 Rock formations on opposite sides of the Atlantic do not match.
NE14 Rock formations on corresponding coasts of the Atlantic would be nearly identical.
E15 Basalt at the ocean bottom is denser and more rigid than deeper basalt.
NE15 The ocean floors are less rigid than the continents.
E16 There are many differences between pre-Cretaceous flora and fauna on different continents.
NE16 Fossil remains of pre-Cretaceous flora and fauna should be identical on corresponding coastlines of the Atlantic.
E17 Islands in the Guinea Gulf are older than the Cretaceous period.
E18 Measurements made in 1926 show no movement of Greenland.
NE18 Measurement would indicate motion of Greenland.

Hypotheses of Wegener:

W1 The crust originally covered the earth, then folded together in the Paleozoic to create Pangaea.
W2 Tidal forces break up the continents and move them westward.
W3 Polfluchtkraft moves the continents toward the equator.
W4 Pangaea broke up in the Cretaceous period.
W5 Continents move through the less rigid ocean floors.
W6 Continents are folded and form mountains due to the resistance of the more rigid ocean floors.
CW1 Large sections of the earth's crust are in gravitational equilibrium (isostasy).

Hypotheses of Contractionists:

C1 The earth has been contracting since birth due to planetesimal settling.
C2 The crust is compressed and fractures into blocks.
C3 Blocks can override one another and tip.
C5 Isthmus connections between continents served as land bridges.
C6 Oceans and continents remain relatively the same.
C7 Continental crust can move vertically but not laterally.

Table 7.4

Explanations and Contradictions for Simulation of Wegener's Opponents

Explanations:

(explain (W3) NE5)
(explain (W5) NE6)
(explain (W1 W2 W3) E4)
(explain (W2 W4 W5) E3)

(explain (W5) NE7)
(explain (W4) NE8)
(explain (W2) NE9)
(explain (W6) NE10)
(explain (W4) NE11)
(explain (W6) NE11)
(explain (W4) NE12)
(explain (W6 W4) NE13)
(explain (W4) NE14)
(explain (W5) NE15)
(explain (W4) NE16)
(explain (W5) NE18)
(explain (W2 W3) W5)
(explain (W2) W4)
(explain (W3 W6) E1)
(explain (W5 CW1) E2)
(explain (C1) C2)
(explain (C1 C3) E1)
(explain (C1 C2) E2)
(explain (C6 C7) E5)
(explain (C1 C2) E6)
(explain (C2 C3) E7)
(explain (C6 C7) E8)
(explain (C6 C7) E9)
(explain (C1 C2 C3) E4)
(explain (C2 C3 C6) E11)
(explain (C6) E12)
(explain (C6 C7) E14)
(explain (C1 C2) E15)
(explain (C6 C7 C5) E16)
(explain (C6 C7) E17)
(explain (C7) E18)

Contradictions:

(contradict W5 W6)
(contradict C3 W6)
(contradict W1 C6)
(contradict W4 C6)
(contradict W5 C7)
(contradict E5 NE5)
(contradict NE6 E6)
(contradict NE7 E7)
(contradict NE8 E8)
(contradict NE9 E9)
(contradict NE10 E10)
(contradict NE11 E11)
(contradict NE12 E12)
(contradict NE13 E13)
(contradict NE14 E14)
(contradict NE15 E15)
(contradict NE16 E16)
(contradict NE18 E18)

Data:

(data (E1 E2 E3 E4 E5 E6 E7 E8 E9 E11 E12 E13 E14 E15 E16 E18 E19 E20))

Table 7.5
Propositions for Simulation of Hess

Evidence:

E1	Many fossil species are divided by water.
E2	There are sedimentary strata at high elevations.
E3	Oceanic earthquakes produce steeply dipping faults.
E4	There are nearly horizontal faults in mountain ranges.
E5	The geology of opposite sides of the Atlantic is similar.
E6	Transoceanic continental margins are congruent.
E7	Scandinavia is rising 1 cm./yr.

Table 7.5 *cont.*

E9	Existence of mountains.
E10	Structure of Alps requires major compression.
E11	Seismic refraction indicates that oceanic crust is very uniform.
E12	There are gravity deficiencies at oceanic trenches.
E13	Deep earthquakes occur near oceanic trenches, at an angle to surface.
E14	Trenches are colder than the rest of the ocean floor.
NE14	Trenches over a cooling mantle should be as warm as the rest of the ocean floor.
E15	Mid-ocean ridges have high heatflow.
E16	There is a mid-Pacific Mesozoic ridge.
E17	There are relatively few volcanic seamounts on the ocean floor.
NE17	An ancient ocean floor should display many volcanic seamounts.
E18	There are no pre-Cretaceous rocks on the ocean floor.
NE18	An ancient ocean floor should include rocks of all epochs.
E19	Apparent polar wander paths indicate that the north magnetic pole was not always where it is today.
E20	Average heatflow in oceans about same as that of continents, despite radioactive rock.
NE20	A cooling mantle would leave the ocean floor cooler than the continents.
E21	There is a relatively thin veneer of sediments on the ocean floor.
NE21	An ancient ocean floor should be covered with a thick layer of sediment.
E22	Apparent polar wander paths differ between continents, but are consistent within continents.
E23	There is a worldwide system of mid-ocean ridges.
E24	Most oceanic earthquakes occur on central rift.
E25	Seismic velocities under the crests of ridges are lower than normal, but become normal again on ridge flanks.

Contractionist hypotheses:

C1	The upper mantle of the earth has been gradually cooling.
C2	The earth has been contracting since birth.
C3	The crust is compressed and fractures into blocks.
C4	Blocks subside and become elevated.
C8	Earth's magnetic pole has wandered over time.
C9	Ocean basins are older than continents.
C10	Ocean basins are parts of the original crust that have been least altered.
C11	Continents grow over former ocean floors, starting with accretion of deltas.
C12	Topography of ocean floors is complex, more rugged than that of continents.

C13	There are large numbers of islands and seamounts on the ocean basin that originated as volcanoes.
C14	After shelf accretion, a conical fracture zone forms further offshore than the shelf, located by earthquake foci.
C15	Volcanism starts as a result of the fracturing and produces a volcanic arc.
C16	Continental material accumulates from volcanic residue and erosion of the continents, creating shallow inland seas.
C17	Further depression, compression, and uplift convert inland seas to land and volcanic arcs into mountain arcs.
C18	Primary arcs of mountain building meeting at an acute angle create a short mountain range of great height.
C19	Measures of the earth's radioactivity are too high.
C20	Ocean trenches are due to overriding by inner blocks of arcs.

Hess's seafloor spreading hypotheses:

CS1	Mantle is viscous.
CS2	Ice sheets are heavy enough to depress continental margins.
S2	The mantle is convecting at the rate of 1 cm./yr.
S4	The continents are carried passively on the mantle at a uniform rate by convection and do not plow through oceanic crust.
S7	Continents were once all connected.
S8	The earth's pole of rotation has moved with respect to the crust.
S10	Rising limbs coming up under continental areas fracture them and move the fragmented parts away from one another.
S11	Mountains form on the leading edges of continents underthrusting one another.
S14	Continental and oceanic crust are very different.
S15	The mantle's convection cells have rising limbs under mid-ocean ridges.
S16	Mantle material comes to the surface at the crest of mid-ocean ridges and becomes oceanic crust.
S17	The uniform thickness of the oceanic crust results from the maximum height that the 500 degree C. isotherm can reach under the mid-oceanic ridge.
S18	Higher temperatures and intense fractures under ridge crests explain lower seismic velocities there.
S19	Mid-ocean ridges are ephemeral features having a life of 200 to 300 million years, the life of a convecting cell.
S20	The whole ocean is virtually swept clean (replaced by new mantle material) every 300 to 400 million years.
S21	The leading edges of continents are strongly deformed when they impinge upon the downward moving limbs of convecting mantle.
S22	The oceanic crust, buckling down into the descending limb, is heated and loses its water to the oceans.

Table 7.6
Explanations and Contradictions for Simulation of Hess

Explanations:

(explain (C1) C2)
(explain (C1) NE20)
(explain (C1) NE14)
(explain (S4 S7 S10) E1)
(explain (C2 C3 C11 C14 C17) E2)
(explain (S2 S4 S11) E2)
(explain (C2 C3 C14) E3)
(explain (S2 S15 S18 S22) E3)
(explain (C2 C3 C14 C15 C16 C17 C18) E4)
(explain (S2 S4 S11 S21) E4)
(explain (S2 S4 S7 S10) E5)
(explain (S2 S4 S7 S10) E6)
(explain (CS2 CS1) E7)
(explain (C2 C3 C4 C17) E7)
(explain (S2 S4 S11) E9)
(explain (C2 C3 C4 C11 C14 C16 C17) E9)
(explain (S2 S4 S11) E10)
(explain (C11 C14 C16 C17 C18) E10)
(explain (S2 S14 S15 S16 S17) E11)
(explain (C9 C10) E11)
(explain (S2 S22) E12)
(explain (C2 C3 C4 C14 C15 C20) E12)
(explain (S2 S21 S22) E13)
(explain (C11 C13 C14) E13)
(explain (S2 S22) E14)
(explain (S2 S15) E15)
(explain (C9 C10 C12 C13) E16)
(explain (S2 S15 S19) E16)
(explain (C12 C13) NE17)
(explain (S2 S15 S16 S20) E17)
(explain (S2 S15 S16 S20) E18)
(explain (C9 C10) NE18)
(explain (S4 S7 S8 S10) E19)
(explain (C8) E19)
(explain (S2 CS1 S15) E20)
(explain (C19) E20)
(explain (S2 S15 S16 S20) E21)
(explain (C9 C10) NE21)
(explain (S4 S7 S8 S10) E22)
(explain (C8) E22)
(explain (S2 S10 S15) E23)
(explain (C2 C3 C12) E23)
(explain (S2 S15 S16 S18) E24)
(explain (S2 S15 S16 S18) E25)

Contradictions:

(contradict E18 NE18)
(contradict E21 NE21)
(contradict E20 NE20)
(contradict E17 NE17)
(contradict E14 NE14)
(contradict C9 S20)
(contradict C10 S16)
(contradict C13 S20)
(contradict C17 S11)
(contradict C18 S21)
(contradict C20 S22)
(contradict C2 S4)

Data:

(data (E1 E2 E3 E4 E5 E6 E7 E9 E10 E11 E12 E13 E14 E15 E16 E17 E18 E19 E20 E21 E22 E23 E24 E25))

CHAPTER 8

Revolutions in Physics

OVER the past five hundred years, physics has had no fewer than four conceptual revolutions. In the sixteenth century, Copernicus developed an astronomical theory that removed the earth from the center of the universe, where the conceptual systems of Aristotle and Ptolemy had placed it. In the seventeenth century, Isaac Newton developed a mechanical theory that embraced both celestial and terrestrial motion and supplanted the vortex theory of Descartes. Newton's views held sway until the twentieth century, when Einstein's theory of relativity offered a radically new way of viewing matter in motion. Also in this century, quantum theory revolutionized how physicists think of light and subatomic particles.

The analyses in this chapter will not be as detailed as the three earlier discussions of revolutions, but will nevertheless show that the revolutions in physics are well understood in terms of conceptual change and explanatory coherence. The Copernican, Newtonian, Einsteinian, and quantum revolutions all involved substantial change in kind-relations, and the latter involved radical change in part-relations as well. In each of the four cases, analysis shows that the new theory had greater explanatory coherence than its predecessor.

8.1 COPERNICUS

When Copernicus published his major astronomical work in 1543, the dominant picture of the universe was still the one that the Greek philosopher Aristotle had sketched in the fourth century B.C. About five hundred years after Aristotle, the Egyptian astronomer Ptolemy developed a mathematical system that made possible detailed predictions of planetary observations. Copernicus's work was intended as an explicit alternative to Ptolemy's. I shall give only a brief summary of the development of Copernican ideas sufficient for the analysis of conceptual change; excellent detailed discussions are available by Boas (1962), Butterfield (1965), Kuhn (1957), and Holton and Brush (1985).

8.1.1 Aristotle

Aristotle's essay *On the Heavens* was the central text on the structure of the heavens during the later Middle Ages. Aristotle's conceptual system was very different from ours, but careful examination of his writings displays how coherently it integrated various phenomena. Figure 8.1 is a sketch of the relations of some of the major concepts in Aristotle's system. Notice first that celestial bodies such as the sun and the planets (not counting earth) are a distinct kind of thing from terrestrial bodies, the physical objects one finds on earth. Terrestrial bodies are composed of mixtures of the four basic elements, earth, water, air, and fire. These bodies are subject to change, but the celestial bodies are unchanging: "For in the whole range of time past, so far as our inherited records reach, no change appears to have taken place either in the whole scheme of the outermost heaven or in any of its proper parts" (Aristotle 1984, 451).

Most important, the celestial bodies are naturally in motion, whereas the nature of terrestrial bodies is to be at rest. Motion and rest can each be either natural or unnatural; if constraint is required to make a body move or stay at rest, then that is an unnatural state for it. "A thing moves naturally to a place in which it rests without constraint, and rests naturally in a place to which it moves without constraint" (Aristotle 1984, 458). For celestial bodies such as the stars, circular motion is the natural state. But for terrestrial bodies, the natural state is rest in the proper place, which varies depending on their constituents. For the elements earth and water, the natural resting place is the center of the earth, which is also the center of the universe. Hence bodies containing earth and water naturally move toward the center of the earth, with ones containing more water resting on top of ones containing more earth. In contrast, the natural movement of air and fire is upward toward the heavens. In all these cases, the natural motion is rectilinear and straight up or down. The natural motion of a terrestrial body is to a place where it can assume its natural state of rest. As Cohen (1980) points out, motion for Aristotle was not a state, but a *process* in which a potentiality achieves an actuality.

On Aristotle's view, the stationary earth is at the center of the universe, surrounded by the spherical heaven. The spherical earth is surrounded by the larger sphere of the heavens, on which rotates the stars which are themselves spherical. Aristotle did not discuss the motions of the planets in *On the Heavens*, but Ptolemy later worked out an intricate system to explain their apparent motion as well. Copernicus, Galileo, and Newton together overturned the entire Aristotelian system. Copernicus rejected the spatial arrangement, removing the earth from the center of the universe, and he altered the kind relations concerning celestial bodies, installing earth as a kind of planet.

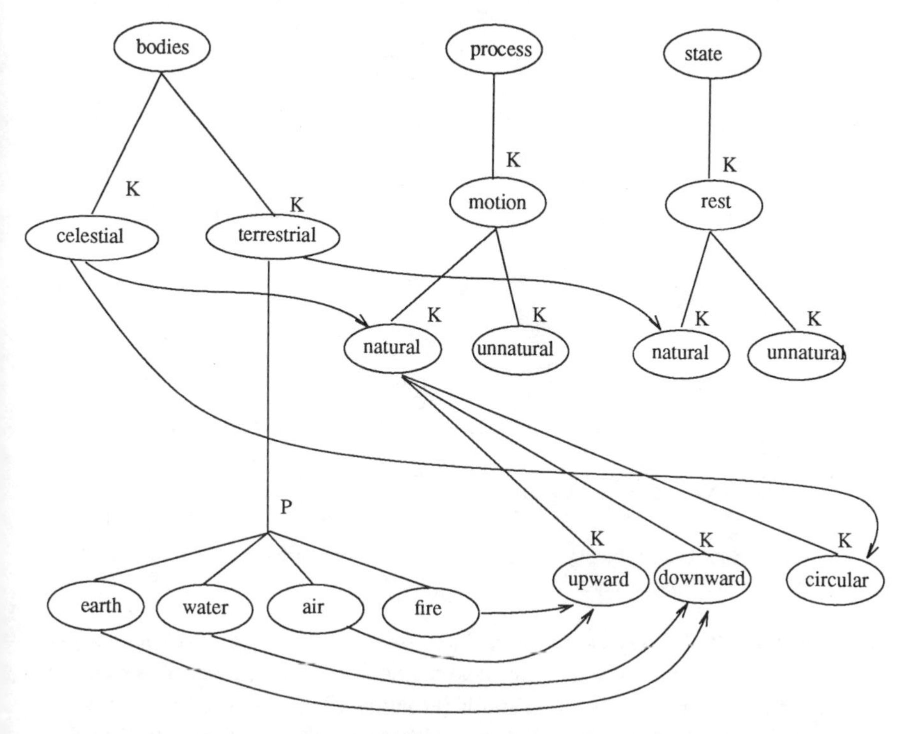

Figure 8.1. Conceptual system of Aristotle. Straight lines indicate kind-relations, except for the part-relation marked "P." Curved lines indicate rules: circular motion is natural for celestial bodies and rest is natural for terrestrial bodies. For terrestrial bodies not at rest, the natural motion of air and fire is upward, and the natural motion of earth and water is downward.

Galileo's principle of inertia abolished the distinction between natural and unnatural motion and rest, and Newton finally eliminated the distinction between celestial and terrestrial bodies, showing them all to be subject to the same laws of motion. But we should appreciate how Aristotle's conceptual organization contributed to explanations of such important facts as the movements of the planets and the behavior of falling bodies.

8.1.2 Ptolemy

Around 150 A.D., Ptolemy's *Almagest* set out a system of the world that was to dominate astronomy for over 1400 years. Ptolemy went far beyond Aristotle's general picture of nested spheres, for he developed the mathematics to account for the detailed motions of the planets. He provided several argu-

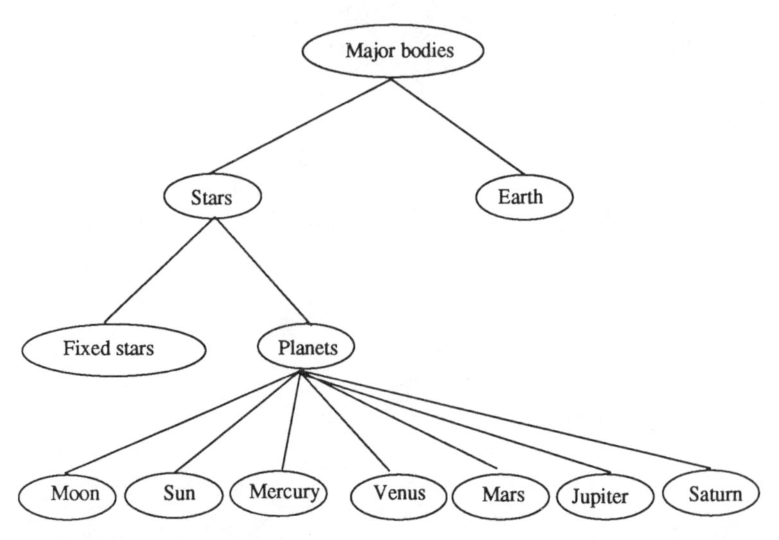

Figure 8.2. Conceptual system of Ptolemy, supplementing Figure 8.1. Straight lines indicate kind-relations.

ments that the heavens are spherical and rotate: this hypothesis explains why sundials work; the motion of the heavenly bodies is free and unhampered; and the sphere is the greatest of all solids (Ptolemy 1984, 39–40). A similar combination of empirical and not-so-empirical arguments supported the conclusion that the earth is spherical, is in the middle of the heavens, and does not move. Most of the volume is a trigonometric tour de force in which Ptolemy shows how to calculate the occurrences of many astronomical phenomena such as eclipses.

To account for the motions of the planets, Ptolemy was forced to deviate from the letter of the Aristotelian description of the universe. Ptolemy tried to maintain the use of uniform circular motions around a stationary earth, but had to introduce several kinds of motion to handle the planets. Ptolemy allowed the planets to move in eccentric paths around the earth, accounting for their seeming to be closer to the earth in some seasons than in others. To explain the retrograde motion of some of the planets, which seem to go forward and then backward in the sky, Ptolemy postulated epicyclic motion in which a planet moves in small circles around points on the large circle that takes it around the earth. Ptolemy also introduced the *equant*, a point offset both from the center of the universe and the now-eccentric earth. So planets were no longer said to revolve with uniform circular motion around the earth, but only with respect to the equant and the planet's epicycles.

The structure of Ptolemy's conceptual system was very different from ours today. Figure 8.2 shows the kind-relations of the astronomical bodies discussed by Ptolemy. Note that the earth is a unique kind, and that, in a usage

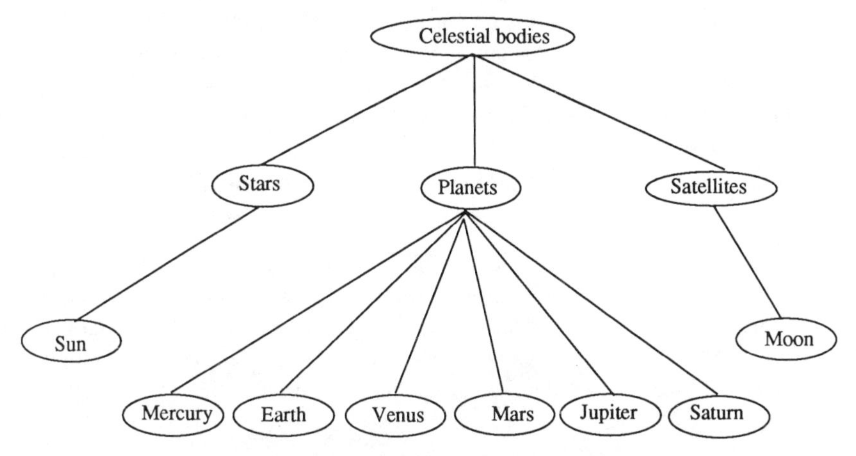

Figure 8.3. Conceptual system of modern astronomy, to be contrasted with Figure 8.2. Straight lines indicate kind-relations.

still found in Galileo (1957), the planets are wandering stars. The moon and sun are both counted as planets. With Copernicus, the earth joins the ranks of planets, and the sun and moon take on a special status. The general concept of a planetary moon arose only with Galileo's discovery of the moons of Jupiter in 1609, and the sun was not recognized as a star until around 1800. Figure 8.3 shows the modern reorganization of the conceptual structure found in Figure 8.2.

Although it suffered from problems such as an inability to account for the perceived size of the moon, the Ptolemaic system was highly coherent both with observation and with other dominant views. Its predictions of the paths of the planets were awkward, but they were the best available until Kepler in 1609 realized that the planetary orbits are ellipses rather than circles, and it did not violate commonsense views that the earth is stationary and the sun and stars move. In addition, Ptolemy's system fitted with Aristotle's physics and his explanation of gravity as due to the movement of things to their natural resting place at the center of the universe. Ptolemy's system thus possessed great explanatory coherence.

8.1.3 The Development of Copernicus's Theory

Copernicus, however, found the Ptolemaic system lacking and sought to develop a superior account. In his major work, *On the Revolutions of the Celestial Orbs*, Copernicus (1978, 4) outlined several sources of dissatisfaction with Ptolemy's account. First, it did not give the motion of the sun and moon exactly enough to establish a constant length for the tropical year. Second, to account for the motions of the planets, Ptolemy and his followers introduced eccentrics and epicycles that contradict the "first principles of uniform motion." Finally, the Ptolemaic system does not fix the relative location of the planets and the structure of the universe. Copernicus reacted to this confusion by reading the works of a wide range of ancient philosophers, and encountered the view of Heraclides and others that the earth moves. He states (Copernicus 1978, 5):

> And even though the idea seemed absurd, nevertheless I knew that others before me had been granted the freedom to imagine any circles whatever for the purpose of explaining the heavenly phenomena. Hence I thought that I too would be readily permitted to ascertain whether explanations sounder than those of my predecessors could be found for the revolutions of the celestial spheres on the assumption of some motion of the earth.

Copernicus adopted the view that the earth moves around 1510 and shortly thereafter wrote a short essay, the *Commentariolus* (Copernicus 1959). This essay contains the essence of the Copernican view that places the sun at the center of the universe with the earth and the other planets revolving around it, but it lacked the detailed mathematical apparatus that Ptolemy provided for calculating astronomical occurrences. Copernicus therefore devoted years to making observations that enabled him to set key parameters and to doing trigonometric calculations of the sort that Ptolemy had offered in the *Almagest*. Most of *On the Revolutions* is trigonometry, as Copernicus tried to account for Ptolemy's phenomena within his heliocentric system.

Although the details of Copernicus's discovery are not known, it seems that it is better characterized as *coherence-driven* than as *data-driven* or *explanation-driven*. To be sure, Copernicus wanted better accounts of the observations of the sun and moon, but most of all he wanted accounts that did not violate Aristotelian principles of uniform circular motion. For Copernicus, the Ptolemaic system had an internal contradiction, exalting the perfect motion of the heavenly spheres while postulating epicycles and other deviations from perfection. Copernicus did not himself generate the hypothesis that the earth revolves around the sun, since he was able to borrow it from the ancients, but he did work out the details that made possible the explanation of numerous observations.

8.1.4 Conceptual Change in Copernicus

As many commentators have pointed out, Copernicus's system was far from a complete break with the Ptolemaic one. In one respect, Copernicus was more Aristotelian than Ptolemy was, since he took so seriously the idea of uniform circular motion. He introduced no new concepts and maintained the central concepts of the Aristotle/Ptolemy system. Although he rejected the analysis of the planets in terms of epicycles and equants, he certainly did not reject these mathematical concepts themselves. Adoption of the Copernican system did, however, require substantial belief revision, in particular rejection of the belief in the stationary earth in favor of the belief that the earth moves, and revision in beliefs about the center of the universe. These beliefs had a large degree of explanatory entrenchment, as defined in section 4.4.

Moreover, revision of these beliefs required much conceptual reorganization. Copernicus's postulation that the earth was a planet, one of the wandering stars, required a major reorganization of the astronomical kind-hierarchy (see Figures 8.2 and 8.3 above, but keep in mind that the second of these is in part post-Galileo). This reorganization is an excellent example of what Chapter 3 called *branch jumping*, since earth jumps from its own branch in the kind-tree to reside under planet, while the sun and the moon jump from the class of planets to other classifications. Thus Copernicus's system involved more than rejection of Ptolemaic beliefs about the earth being stationary and at the center of the universe. Alterations in the kind-hierarchy amounted to substantial conceptual change.

Copernicus was still operating with the Aristotelian concepts of natural and unnatural motion. He said, for example, that rotation is the natural motion of a sphere, so that it is natural for the earth to rotate. He countered Ptolemy's argument that movement of the earth would be destructive for everything on it by saying that motion so natural cannot be destructive (Copernicus 1978, 15). In contrast, the immensity of the heavens makes it less natural for them to revolve. Copernicus also had to deal with the problem of gravity, since Aristotle's concept of gravity as natural movement toward the center of the universe no longer applies when the earth is moved from the center. Copernicus instead suggested that "gravity is nothing but a certain natural desire, which the divine providence of the Creator of all things has implanted in parts, to gather as a unity and a whole by combining in the form of a globe" (1978, 18). This impulse is found not just in the earth, but in the sun, moon, and other planets. The history of the concept of gravity is fascinating; we shall return to it in later sections on Newtonian mechanics and relativity theory.

Copernicus's system thus requires substantial belief revision and conceptual reorganization compared to Ptolemy's. The abandonment of Ptolemy's theory was justified by the Copernican system's greater explanatory coherence.

8.1.5 The Explanatory Coherence of Copernicus's Theory

Nowak and Thagard (1991a) have done a detailed ECHO analysis of the comparative explanatory coherence of the Copernican and Ptolemaic systems, based on the original texts of *On the Revolutions* and the *Almagest*. I shall not, however, repeat the analysis here, since the simulation is large and complicated, requiring more than 140 propositions, 60 of which represent the astronomical evidence that the two systems explain. Copernicus modeled his book on Ptolemy's, and they both can be understood as providing explanations of the observed motions of: (1) the superior planets, Mars, Jupiter, and Saturn, which are outside the sun's orbit for Ptolemy and the earth's orbit for Copernicus; (2) the inferior planets, Mercury and Venus, which are inside the earth's orbit for Copernicus; (3) the moon; and (4) the sun and stars. Copernicus's most important hypotheses were that the sun is immobile at the center of the universe and that the earth revolves around the sun once a year, uniformly in a circle. These hypotheses contradicted Ptolemy's central hypothesis that the earth is always at the center of the heavenly sphere. Based on the input we gave it, ECHO finds the Copernican position much more coherent than the Ptolemaic, not surprising since we were modeling Copernicus's view. Copernicus dominates Ptolemy for four reasons, although computational experiments revealed that the last two are not very important in the simulation. First, Copernicus explained more than Ptolemy, who did not explain why the moon's size and parallax do not change visibly, why Mercury and Venus never appear in opposition to (across the sky from) the sun, or why Mars, Jupiter, and Saturn often appear in opposition. Second, the system of Copernican hypotheses is simpler than the Ptolemaic set; our ECHO analysis identified 29 hypotheses for Copernicus and 39 for Ptolemy. The reason for the difference is that Copernicus had a much more economical way of explaining the motions of the superior and inferior planets. Ptolemy, in contrast, had to make different assumptions about the epicycles of the superior planets and the epicycles of the inferior planets. At the level of qualitative explanations, Copernicus is clearly superior, since he explains more with less. Kuhn (1957), however, points out that Copernicus found it necessary to complicate his system with additional circles and motions in order to predict the position of planets with accuracy comparable to Ptolemy's system. Nevertheless, Copernicus held that his view was to be preferred for its simplicity:

> We therefore assert that the center of the earth, carrying the Moon's path, passes in a great circuit among the other planets in an annual revolution round the Sun; that near the Sun is the center of the universe; and that whereas the Sun is at rest, any apparent motion of the Sun can be better explained by motion of the Earth. . . . I think it easier to believe this than to confuse the issue by assuming a vast number of Spheres, which those who keep earth at the center must do. We thus rather follow Nature, who producing nothing vain or superfluous often prefers to endow one cause with many effects. (quoted in Kuhn 1957, 179)

There are two additional minor reasons why ECHO prefers Copernican hypotheses to Ptolemaic ones. The third reason, in addition to the major ones of explanatory range and simplicity, is that Ptolemy's hypotheses do not simply fail to explain some facts about the moon, but imply incorrectly that the moon's size and parallax should change visibly in the course of its orbit (negative evidence). The fourth reason is that, on Copernicus's account, Ptolemaic epicycles contradict the central principle that the motions of heavenly bodies are uniform and circular. A logician might exclaim: "If the Ptolemaic system really is self-contradictory, that would be grounds for rejecting it immediately, so explanatory coherence is irrelevant." But there is much more to belief revision in complex systems than a simple logical model could hope to handle. ECHO treats the contradiction in the Ptolemaic system as an internal tension that counts against it but that is not in itself devastating.

If Copernicus's system really was more worthy of acceptance than Ptolemy's, why did it take more than 100 years for it to be adopted? One sociological reason was the opposition of the Catholic Church, which declared the heliocentric universe heretical, eventually prosecuting Galileo for his defense of it. But the slowness of acceptance of the Copernican system also can be understood in cognitive, explanatory-coherence terms. We saw that the substantial advantage in qualitative explanations of general phenomena possessed by Copernicus did not carry over to quantitative explanations of particular observations. Only when Kepler discovered that planets move in ellipses did the predictive accuracy of the heliocentric system become adequate without proliferation of circles. In addition, some astronomers found unconvincing Copernicus's attempt to explain the absence of stellar parallax. Another limitation on the coherence of the Copernican system was that its central hypothesis asserted that the earth moves, contradicting the accepted propositions of Aristotelian physics and everyday observation. The Copernican system was a challenge to more than just Ptolemaic astronomy, calling into question the Aristotelian explanation of gravity as tendency to move to the center of the earth. An ECHO analysis of the full conceptual system of the sixteenth century would require confronting Copernicus propositions with a tag-team of Ptolemy and Aristotle. Only when Galileo joined the fray with his challenge to traditional terrestrial physics could the ancient system be challenged as a whole.

What is the relation between the Copernican and Ptolemaic theories? Even though Copernicus did not introduce any new concepts, he did initiate a reorganization of the kind-relations involving the earth, and he rejected the key Ptolemaic tenets concerning our planet's immobility and centrality. Hence the Copernican system is better described as potentially *supplanting* the Ptolemaic system than as incorporating or sublating it. Historically, however, the Ptolemaic/Artistotelian system was not completely routed until Newton produced his grand synthesis of celestial and planetary motion.

8.2 NEWTON

Isaac Newton was born in 1642, almost a hundred years after Copernicus's death. Newton's *Principia*, published in 1687, used three basic laws of motion to explain a very wide range of terrestrial and celestial phenomena, thus tying together and solidifying the insights of Copernicus, Galileo, Kepler, and Descartes. Appreciation of Newton's theory requires understanding of four concepts integral to Newton's explanations: inertia, mass, force, and gravity. In all four cases, Newton took existing concepts, modified them substantially, and melded them into a theory of unprecedented explanatory power. This section will trace out the conceptual changes that Newton produced and discuss the explanatory coherence of his theory, which Newton compared with Descartes's theory of vortex motion.

8.2.1 Newton's Conceptual Changes

Newton did not leave much trace of the development of his theory, so it is impossible to reconstruct it in enough detail to model it computationally. He appears to have arrived at many of the key mathematical ideas of his system in the mid-1660s, when he developed the differential calculus and deduced that the force keeping the planets and the moon in their orbits is inversely proportional to the square of the distance between them and the centers around which they revolve. He returned to the examination of astronomical questions in 1684, prompted by a query from Edmund Halley. Eighteen months of intense labor produced Newton's landmark theory and the book, *Philosophiae Naturalis Principia Mathematica*, that presented it.

Newton presented his theory in Euclidean form, with definitions and axiom-like laws. The conceptual novelties of the Newtonian system are embedded in the three famous laws, the first of which is:

> Every body continues in a state of rest, or of uniform motion in a right line, unless it is compelled to change that state by forces impressed upon it. (Newton 1934, vol. 1, 13)

This law embodies a concept of inertia, which Newton earlier (p. 2) defined as "a power of resisting, by which every body, as much as in it lies, continues in its present state, whether it be of rest, or of moving uniformly forwards in a given line." He also described inertia as an "innate force of matter." Newton's first law and the concept of inertia dramatically challenge the Aristotelian view of motion discussed in section 8.1.1, according to which the natural state for celestial bodies is motion and the natural state for terrestrial bodies is rest. For Newton, motion and rest are equally natural, and there is no distinction made between celestial and terrestrial bodies. Newton thus offered a conceptual organization very different from that shown in Figure 8.1, collapsing distinctions that were central to the Aristotelian system. The concepts of natural and unnatural motion are deleted, and the division of bodies into two subkinds, celestial and terrestrial, is abandoned for physical purposes.

Compare Darwin's collapse of the distinction between species and varieties described in section 6.2.2.

Newton did not form this concept of inertia himself, but borrowed it from Descartes's *Principles of Philosophy*, published in 1644. Descartes had stated as the first law of nature that "every thing, in so far as it can, always continues in the same state; and thus what is at once in motion always continues to move" (Descartes 1985, vol. 1, 240). Shortly before, Galileo had come close to the modern concept of inertia, but his system retained elements of the *impetus* theory that medieval thinkers such as Buridan had proposed to overcome the difficulties of explaining some kinds of motion within the Aristotelian system. For Aristotle, it was a problem to explain why a thrown object should continue to move after it left the thrower's hand. According to impetus theory, throwing the object imparted impetus to it, making the object continue to move until the impetus expired. Galileo can be interpreted as viewing the planets as moving because of a kind of circular impetus, rather than because of Newton's rectilinear inertia (Shapere 1974).

Newton's second law is:

> The change of motion is proportional to the motive force impressed; and is made in the direction of the right line in which that force is impressed. (Newton 1934, vol. 1, 13)

This statement differs from the modern statement of Newton's second law, $F = ma$: force equals mass times acceleration. Newton defined "quantity of motion" as "arising from the velocity and quantity of matter conjointly" (p. 1). In modern terms, change of motion is change in momentum, where momentum is defined as mass times velocity, mv.

Newton's tersely stated law conceals major innovations in two key concepts, *mass* and *force*. Newton was one of the first to distinguish mass, or quantity of matter as he also termed it, from *weight*. Using the terminology of Chapter 3, the conceptual change involved here is best described as differentiation, since a vague single concept is divided into two. The differentiation makes possible understanding of the weight of an object as the combined result of its mass and the force of gravity.

Newton distinguished between two kinds of force, impressed and centripetal. An impressed force is an action exerted on a body to change its state. A centripetal force, in contrast, draws bodies "towards a point as to a centre" (p. 2). Before Newton, the primary meaning of force was the first of these, consistent with the everyday notion of force. Newton added the second kind of force, and reconceived gravity as a kind of centripetal force. On this view, the motions of the planets can be understood in terms of the first and second laws: planetary motion is a combination of inertial motion and gravitational (centripetal) force drawing the planets to the sun. The explanations of planetary motions offered by Newton in terms of his laws depended, therefore, on several sorts of conceptual change: differentiating mass from weight, adding a new kind of force, and recategorizing gravity as a kind of centripetal force. Even more fundamentally, it required abandonment of the Aristotelian distinction between motion as a process and rest as a state (Cohen 1980, 182). For Newton, both of these are kinds of state subject to the first law concerning inertia. The reclassification of motion as a state is a clear case of branch jumping.

For Aristotle, gravity was an intrinsic property of a body, but for Newton it became a force exerted on a body by another body. This change is one of the most important instances of branch jumping in the history of science. Reconceiving gravity as a force made possible the great synthesis of planetary and terrestrial phenomena that Newton accomplished. That bodies fall on Earth was attributed to the centripetal force gravity, the same cause as the circular motions of the planets. Figure 8.4, adapted from the insightful analysis of Nersessian (1989), sketches the conceptual organization. Planetary motion, as a case of accelerated circular motion, is naturally explained in terms of a combination of gravity and inertia. By contrasting Figure 8.4 with Figure 8.1, the reader should be able to appreciate the major conceptual reorganization that Newton accomplished, building on Copernicus, Galileo, and Descartes. Newton collapsed distinctions that were important for Aristotle, introduced subkinds not found in the Aristotelian system, and reclassified motion. Planetary motion for Aristotle was the eternally circular kind natural for all celes-

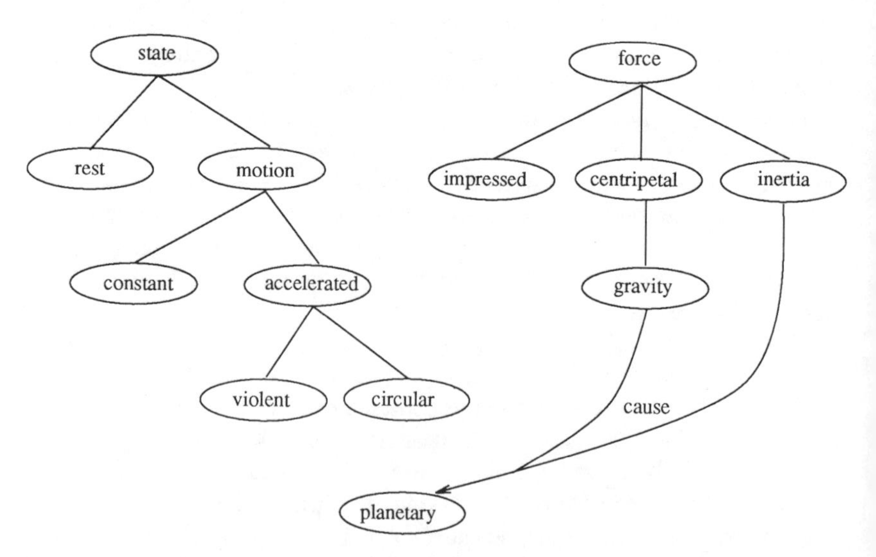

Figure 8.4. Conceptual system of Newton, to be contrasted with Figure 8.1. Straight lines indicate kind-relations. The curved lines represent the rule that inertia and gravity together cause circular planetary motion.

tial matter, but Newton explained planetary motion as the result of a combination of gravitational force and inertia.

The relational nature of the concept of force is brought home in Newton's third law:

> To every action there is always opposed an equal reaction. (Newton 1934, vol.1, 13)

According to Cohen (1980, 177), this law arose as a generalization of cases of impact, where colliding bodies exert force on each other, but its extension to the force of gravity was radical. From the Aristotelian perspective, the thought that a body could exert a force on the earth was unthinkable. A major challenge to the computational approach to understanding the development of scientific knowledge would be to model the discovery of Newton's second and third laws. They obviously were not directly data-driven, since there is much more to them than simple generalizations about observed results. But they are also not easily classified as explanation-driven, since there are no specifiable observations that Newton intended them to explain. Rather, their development seems to be part of general reasoning patterns aimed at creating a system of the world capable of explaining a great many facts. Cohen (1980) provides an interesting discussion of the development of Newton's third law, but without nearly enough detail to permit speculation about the cognitive processes involved in Newton's creative leaps.

Newton's concepts of space and time did not challenge Aristotelian concepts, but they must be mentioned here because of their relevance to the discussion of relativity theory in section 8.3. Newton assumed that concepts of time and space were "well known to all." But he rejected the conception of the common people of space and time in terms of relations to sensible objects. He asserted (Newton 1934, vol. 1, 6): "Absolute, true, and mathematical time, of itself, and from its own nature, flows equably without relation to anything external. . . . Absolute space, in its own nature, without relation to anything external, remains always similar and immovable." We shall see how relativity theory required abandonment of these concepts of absolute time and space.

8.2.2 The Explanatory Coherence of Newtonian Mechanics

Although Newton's theory met with stiff resistance from some circles, particularly French followers of Descartes, it came to dominate physics in the eighteenth and nineteenth centuries. Newton was able to use his three laws to explain a vast array of terrestrial and celestial phenomena. A full analysis of the explanatory coherence of Newton's system will not be presented here, since it requires more than 150 propositions (Nowak and Thagard 1991b). I shall merely summarize the explanatory structure of Newtonian mechanics, indicating how it provides an excellent alternative to Cartesian physics as well as to the Aristotelian system.

It is important not to treat Newton's principles just as a set of axioms from which a set of theorems can be deduced, without paying attention to the different ways laws are used in various explanations. Some of his explanations of terrestrial phenomena involve only one of the three laws, whereas others require pairs of them. For example, Newton cited as evidence for the third law involving action and reaction that if you press a stone with your finger, the finger is also pressed by the stone. Galileo's discovery that the descent of bodies varies as the square of time is explained using the first two laws, invoking both inertia and force. Newton's explanations also involve many levels. His explanations are like Darwin's and unlike Lavoisier's, in that they involve hypotheses explaining other hypotheses that explain pieces of evidence. The layering is most striking in Newton's explanations of the motions of the planets, where he used gravity to explain astronomical facts such as Kepler's laws.

Newton did not directly challenge Aristotelian physics, but took for granted the critiques of it by Galileo and Descartes. His presentation in the *Principia* is comparative, however, since he devotes considerable space to criticizing the Cartesian alternative to his views. Several decades before Newton developed his theory, Descartes claimed to be able to explain the motions of the planets by assuming that celestial matter turns continuously like a vortex with the sun at the center. He compared the celestial vortices to whirlpools in a river, with the planets carried around like flotsam in the whirlpool. The bits near the center of the whirlpool complete a revolution more quickly than the ones farther out. "We can without any difficulty imagine all this happening in the same way in the case of the planets, and this single account explains all the planetary movements that we observe" (Descartes 1985, 254).

The ECHO analysis of Newton and Descartes brought to light four major advantages of the Newtonian system over the Cartesian. First, Newton explained numerous facts that Descartes did not. These included the law of Galileo concerning the descent of bodies, and Kepler's third law of planetary motion, that the periods of the orbits of planets are proportional to the 3/2 power of their distances from the sun. Strictly speaking, Newton corrected Kepler's laws while explaining them, since universal gravitation implies that the motions of planets will not be perfectly elliptical, because gravitational forces exerted by other planets have an effect in addition to the sun's attraction. Figure 8.5 presents a small fragment of the explanatory breadth of Newton's principles, showing that the first and second laws, along with the assumption of gravitational force, serve to explain both falling bodies on earth and Kepler's laws of planetary motions (for much more detail, see Nowak and Thagard 1991b). Descartes's vortex theory gave qualitative explanations of planetary motion and falling bodies, since a vortex can carry objects both

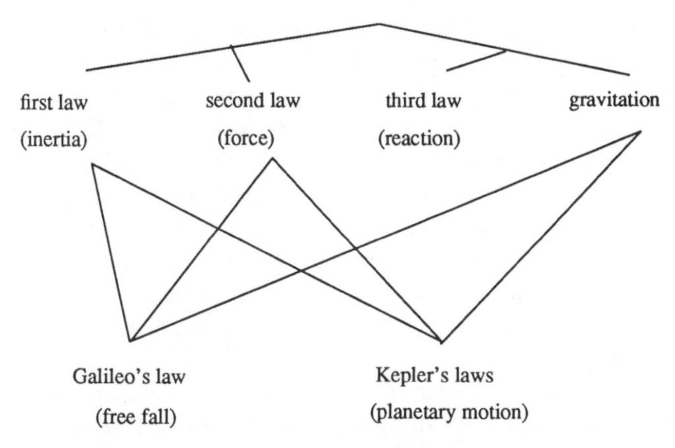

Figure 8.5. Aspects of the explanatory coherence of Newton's theory. Straight lines indicate explanations. The first and second laws and the hypothesis of gravitational force all cohere with each other as the result of these explanations. The third law is used in different explanations (Nowak and Thagard 1991b).

around and toward its center; but Newton explained numerical laws beyond the reach of the vortex theory.

Second, Newton showed that there was much negative evidence for the Cartesian system, which predicted nonexistent terrestrial and celestial phenomena. In particular, Newton demonstrated numerous problems with the vortex theory, showing that it could not account for Kepler's third law.

The third advantage of Newton over Descartes was the greater simplicity of his system. Descartes tended to propose new principles for almost every phenomenon he wanted to explain. Our analysis identified around 70 principles used in explaining only about 20 pieces of evidence. Newton in contrast needed fewer than 50 principles to explain more than 30 pieces of evidence. Newton's explanations were thus simpler than Descartes's, in the sense of simplicity used in the theory of explanatory coherence. Newton's theory is more coherent in a fourth respect, since most of his hypotheses are explained by other hypotheses, with the three laws and the assumption of gravitational force generating many corollaries that are then used to explain evidence. Hence Newton's theory is much more internally coherent than the grab bag of Cartesian hypotheses.

By explaining both terrestrial and celestial hypotheses with a unified set of laws and assumptions, Newton laid to rest Aristotelian physics. The battle with the Cartesians had dimensions that I have not mentioned, such as Descartes's claims that his principles were necessarily true and the arguments of his followers that Newtonian gravitational force was a metaphysically illegitimate postulation. But Newton completed the supplantation of Aristotelian concepts and hypotheses begun by Copernicus and Galileo. We saw that the Newtonian concepts of inertia and gravity are dramatically different from the Aristotelian ones. Although Newton's views were closer to Descartes's, who had acquired a modern concept of inertia, we should probably say that Newton supplanted Descartes as well. Newton completely rejected the Cartesian postulation of vortices whirling the planets around, and the concept of vortex ceased to have any role in the explanation of the solar system. But Newton's system was more radical with respect to the Aristotelian system than it was with respect to the Cartesian system that had learned much from Copernicus and Galileo.

The explanatory coherence of the Newtonian system increased steadily during the two centuries following the publication of the *Principia.* For example, astronomers used Newton's laws to explain away observed anomalies in the orbit of the planet Uranus, postulating the quickly confirmed existence of Neptune. In the twentieth century, however, Newtonian mechanics was challenged by relativity theory and quantum theory.

8.3 EINSTEIN

8.3.1 The Development of Relativity Theory

Newtonian mechanics remained central to physics through the eighteenth and nineteenth centuries, but the latter saw the development of the concept of a *field*, which eventually led to relativity theory and revision of many Newtonian assumptions. In the 1840s, Michael Faraday explained magneto-electric induction by supposing that the space surrounding a current is filled with closed curves of magnetic force. These lines of force were taken to be elastic strains in *aether*, a fluid of the sort that had been postulated to explain the propagation of light. By analogy with water waves, the ancient Greeks had come up with a wave theory of sound that explained how sounds propagate through air and other media. Similarly, Huygens, Fresnel, and Young developed a wave theory of light that supplanted Newton's particle theory of light because it explained various light phenomena such as diffraction. The aether was postulated to provide a medium for light waves to propagate through space.

In 1861, James Clerk Maxwell developed a mathematical theory of electromagnetic fields that showed how to unify electric and magnetic phenomena. In a mechanical spirit, he interpreted his famous field equations in terms of vortices in an aether. Maxwell's theory produced the kind of conceptual change that Chapter 3 called coalescence, since electricity and magnetism were seen to be fundamentally the same kind of thing: mechanical processes in the same sort of aether. Physicists accordingly set out to do experiments to detect the aether. After several flawed attempts, Michelson and Morley in 1887 conducted a careful experiment to detect the relative velocity of the Earth past the aether. Light traveling in the same direction as the earth should have the velocity $c + v$, where c is the speed of light and v is the velocity of the earth through the aether. The experimenters set up an ingenious apparatus that used mirrors to reflect light perpendicular to the motion of the earth and to determine whether it was out of phase with light that had been reflected parallel to the motion of the earth. To their surprise, the two light beams were not out of phase, so that the experiment failed to show the motion of the earth relative to the aether.

In 1889 FitzGerald suggested that the negative result could be explained by hypothesizing *contraction* of the arm of the instrument that was supposed to detect the displacement of the two light beams. Contraction by a factor of

$$\sqrt{1-v^2/c^2}$$

in the direction of motion would suffice to account for the failure of the Michelson-Morley experiment. Lorentz proposed the same hypothesis a few years later as part of a general theory of electrodynamics. Many researchers, however, thought that the FitzGerald/Lorentz contraction was suspiciously ad hoc, contrived merely to handle the Michelson-Morley results.

In 1905, when he was twenty-six years old, Albert Einstein developed the special theory of relativity. Although textbooks sometimes describe this theory as aimed at explaining the Michelson-Morley result, Einstein's own writings show that he was much more motivated by conceptual considerations (Holton 1973). His initial paper, "On the Electrodynamics of Moving Bodies," begins by discussing asymmetries in the applications of Maxwell's equations to the reciprocal action of a magnet and a conductor. If the magnet is in motion and the conductor is at rest, then an electric field arises, but not if the magnet is at rest and the conductor is in motion. Einstein wrote (1952, 37f.):

> Examples of this sort, together with the unsuccessful attempts to discover any motion of the earth relatively to the "light medium," suggest that the phenomena of electrodynamics as well as of mechanics possess no properties corresponding to the idea of absolute rest. They suggest rather that . . . the same laws of electrodynamics and optics will be valid for all frames of reference for which the equations of mechanics hold good.

Einstein called the conjecture about the validity of laws for all frames of reference the Principle of Relativity. The other central principle of the special theory of relativity is that "light is always propagated in empty space with a definite velocity c which is independent of the state of motion of the emitting body" (Einstein 1952, 38). Together, these two principles reject the Newtonian notion of absolute space and render the aether superfluous.

In Chapter 3, I distinguished between data-driven, explanation-driven, and coherence-driven discovery. The development of the special theory of relativity, like that of the general theory described below, clearly was coherence-driven. In his autobiography, Einstein said that the theory resulted from a paradox that he had constructed at the age of sixteen about what would happen if he were to pursue a beam of light with the velocity *c* (Einstein 1949, 53). Intuitively, he would expect to perceive the beam of light as at rest, since he would be traveling at the same speed, just as two cars on the highway at the same speed seem to each other not to be moving. But Maxwell's equations did not allow the possibility of an electromagnetic field at rest. The key to resolving the tension between Maxwell's laws and the thought experiment was to reject the standard concept of simultaneity that presupposes a concept of absolute time. On the new account, every reference body has its own particular time, so nothing is absolutely at rest.

Einstein derived the celebrated relation between mass and energy, popularly summarized in the equation $e = mc^2$, in another paper published in 1905. He deduced from the principle of relativity and the constancy of the speed of light, plus some approximations, that changes in the energy of a body correspond to changes in its mass. Since *c* is so large, 2.998×10^8 meters per second in a vacuum, convertibility of mass and energy is unnoticeable in everyday life. But once a body has a velocity of about twenty percent of the speed of light (more than 37,000 miles per second), the attendant increase in energy corresponds to an increase in mass. As mass increases, it takes more and more force to accelerate the body further, so nothing can go faster than the speed of light. Note that $e = mc^2$ and the constancy of the speed of light are not generalized from observations. Rather, the constancy principle was proposed along with the principle of relativity to resolve incoherencies in existing theories, and the convertibility of mass and energy was derived from them. Nor were these principles generated abductively, to provide some explanations of events that were viewed as puzzling, although we will see that the principles turned out to have considerable explanatory power.

Einstein's original formulation of the relativity principle only implied the invariance of the laws of physics with respect to the Lorentz transformations that incorporated the FitzGerald contraction. Accordingly, this early formulation is now called the *special* principle of relativity, and in combination with the principle of the constancy of the speed of light it constitutes the special theory of relativity. After years of work to bring gravitation within the scope of relativity theory, Einstein in 1915 worked out the mathematics that made possible the generalization for bodies subject to all kinds of motion, including gravitational acceleration. Once again, his discovery is better described as coherence-driven than as data-driven or explanation-driven, since he was attempting to combine considerations from special relativity that inert mass increases with energy with experimental results that showed that gravitational mass is equal to inert mass (Einstein 1949, 65). Using new ideas about accelerated reference systems and nonlinear transformations, Einstein produced the general theory of relativity subsuming the special theory. The general theory increased the explanatory coherence of the whole package because the incorporation of gravitation made possible the explanation of additional phenomena.

My account of the development of Einstein's theory of relativity has been historically sketchy and technically simplified, but many comprehensive descriptions are available (e.g. Pais 1982; Einstein 1961; Holton and Brush 1985; Friedman 1983). Now let us look at relativity theory from the perspectives of conceptual change and explanatory coherence.

8.3.2 Conceptual Change in Relativity Theory

From one perspective, the development of relativity theory does not appear very revolutionary. As Einstein pointed out (1961, 44), the special theory of relativity requires only a slight modification in Newtonian mechanics, since relativistic effects only show up in bodies that are moving very rapidly. The FitzGerald/Lorentz contraction factor

$$\sqrt{1-v^2/c^2}$$

reduces to 1 unless the velocity of the body is high relative to the speed of light. Writing $e = mc^2$ as $m = e/c^2$ highlights how the mass of a body is not notably affected unless its velocity and hence its energy is very high. So it is often said that Newtonian mechanics describes the special case of relativity theory applying to low velocities. Similarly, the general theory of relativity yields approximately the same predictions as Newtonian mechanics for slow-moving bodies and weak gravitational fields. If relativity theory effectively incorporates classical mechanics, what was revolutionary about its adoption?

But there was more to pre-relativity physics than Newton's three laws. Newton explicitly assumed concepts of absolute space and time that relativity theory calls into question. Although Maxwell's equations survive in relativity theory, the nineteenth-century assumptions about the role of the aether in the equations' application were abandoned as superfluous. Einstein's revolution thus involved rejection of at least the following beliefs (Einstein 1952, 1961), the first three of which are eliminated by the special theory alone:

1. Time and space are absolute.
2. There is a luminiferous aether.
3. Objects have no maximum velocity.
4. Euclidean geometry adequately describes space.
5. There are instantaneous gravitational effects.
6. Light travels through space in straight lines.

Thus Einstein's views are certainly revolutionary in that they lead to the rejection of important beliefs.

They also lead to considerable conceptual reorganization. Most obviously, the concept of an aether disappears and plays no role in the new conceptual framework. Einstein also stressed that abandoning the standard concept of *simultaneity*, of events happening at the same time in an absolute sense, was crucial to replacing the old, incoherent views. Shortly after Einstein developed the special theory, Minkowski interpreted it in terms of a four-dimensional space-time continuum, and this reinterpretation was a major contribution to the development of the general theory. Instead of the everyday notions of space and time incorporated into Newton's system, Einstein came to think of time as like the three spatial dimensions. Space and time were no longer conceived as separate from each other. Newton thought that temporal concepts such as duration and simultaneity were unproblematic, but Einstein argued that whether two events are perceived as simultaneous depends on the reference frame of the observer.

Implicitly, this change brings about an enormous modification of our notions of part and whole. Many part-relations are based on spatial relations (a finger is part of a hand) or on temporal ones (a day is part of a week). In ordinary thinking, these two kinds of part-whole relations are entirely distinct. But with the notion of the space-time continuum essential to general relativity there are no longer merely spatial parts or merely temporal parts. One entity is part of another only if it is included in the same region of the space-time continuum. Thus relativity theory brings with it a radical change in the concept of part, as massive a change as Darwin's alteration of the concept of kind. It therefore changes the nature of the part-hierarchy in the fundamental way that Chapter 3 described as tree switching. That is, with relativity theory we find the nature of the tree of concepts established by part-relations changing because the very notion of part is reconceptualized in terms of a space-time continuum. We should not think of space and time as each parts of space-time, because they have no meaning independent of space-time.

The concept of mass underwent a very substantial change also, involving both a coalescence and a differentiation. The convertibility of mass and energy indicated by $e = mc^2$ shows that mass and energy are just two manifestations of mass-energy. For Newton, mass was an absolute notion, the quantity of matter. Concepts of energy developed in the nineteenth century, but it was Einstein who claimed that mass should be eliminated as a separate concept (Einstein 1949, 61). Whereas in nineteenth-century physics there were two independent conservation principles, with Einstein the principles of the conservation of mass and energy become fused into a single principle.

At the same time as the concept of mass was being coalesced with energy, it was also being differentiated. The distinction is now made between the *rest*

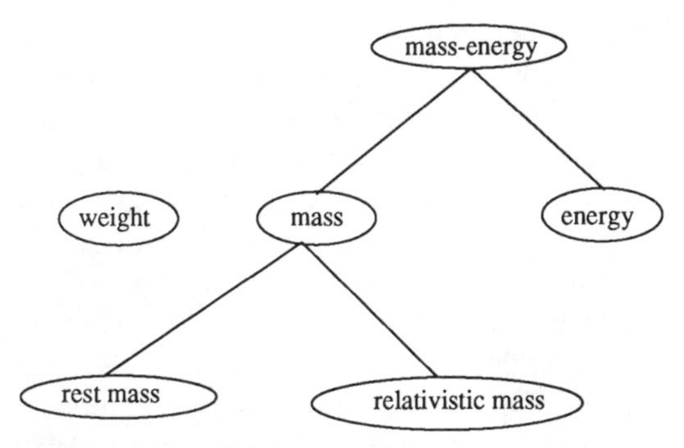

Figure 8.6. The concept of mass in relativity theory. Straight lines indicate kind-relations.

mass of a body when it is not moving relative to an observer, and the *relativistic* mass measured while it moves with a certain velocity relative to an observer. If the velocity is low, rest and relativistic mass are virtually the same, but not so if velocity starts to approach the speed of light. The place of the concept of mass in the new kind-hierarchy is thus shown in Figure 8.6, with mass having acquired a new coordinate concept energy and gained two subordinates. The figure also includes the concept of weight as a reminder of Newton's differentiation of mass and weight discussed in section 8.2.1.

8.3.3 The Explanatory Coherence of Relativity Theory

The adoption of Einstein's special and general theories of relativity involved massive conceptual change and rejection of old assumptions. Yet, for the special theory, at least, the acceptance of Einstein's ideas took only a few years. What made Einstein's theory acceptable over the previous combination of principles drawn from Newton, Maxwell, and Lorentz? Explanatory coherence theory naturally accounts for the superiority of Einstein's views.

In fact, Einstein's own methodological pronouncements fit well with TEC, the theory of explanatory coherence stated in Chapter 4. He wrote (Einstein 1949, 33): "A theory is the more impressive the greater the simplicity of its premises is, the more different kinds of things it relates, and the more extended is its area of applicability." The last consideration, area of applicability, is captured in TEC by clause 2(a) of the explanation principle that implies that the more a theory explains the more coherent it is with the data. The second consideration, concerning relating many different kinds of things, is derivative from the last, since normally a theory gains wide applicability simply by being applied to many different kinds of things. In his first clause, Einstein mentions wanting simple premises, but this quote gives no direct indication that he has in mind the notion of simplicity embodied in TEC. In other places, however, he shows a clear concern with reducing the number of hypotheses. For example, he praises the special theory of relativity over the electrodynamics and optics produced by Maxwell and Lorentz because "it has considerably reduced the number of independent hypotheses forming the basis of theory" (Einstein 1961, 44). Similarly, he criticizes classical mechanics for being able to explain the deviation of the orbit of Mercury from the ellipse predicted by Newton's laws only on the assumption of hypotheses that were devised solely for this purpose (Einstein 1961, 103). Most important, he contrasted the status of the FitzGerald-Lorentz contraction in special relativity with its status in Lorentz's theory: "The theory of relativity leads to the same law of motion, without requiring any special hypothesis whatsoever as to the structure and the behaviour of the electron" (Einstein 1961, 51). In sum, "The relativity theory arose from necessity, from serious and deep contradictions in the old theory from which there seemed no escape. The strength of the new theory lies in the consistency and simplicity with which it solves all these difficulties using only a few very convincing assumptions" (Einstein and Infeld 1938, 192).

Let us now look at the explanatory coherence of relativity theory in greater detail. There is no single text where Einstein argues for his theory against the older ideas, so I shall not attempt a detailed ECHO analysis. But it is important to see how well relativity theory coheres both internally and externally, with the data. What follows is based largely on Einstein (1961). I have already described the key principles of special relativity and it should suffice to see them in relation to each other and to the evidence. Figure 8.7 provides a schematic view of the coherence of Einstein's views around 1920.

In Einstein's case, the explanatory relations are deductive, often being purely mathematical derivations. The special theory of relativity consists essentially of the principle of special relativity and the principle of the constancy of the speed of light. From these two principles Einstein deduced both the Lorentz transformation which explains the negative result of the Michelson-Morley experiment, and the convertibility of mass and energy which explains the nuclear transmutations observed by Rutherford in 1919. In addition the principles explain the results of experiments by Fizeau concerning the speed of light in a flowing liquid (Einstein 1961, 39f.). Thus the two principles provide a unified explanation of three important phenomena. General relativity gets into the picture as a generalization of the special principle, which holds for a smaller class of transformations (Einstein 1949, 69). General relativity explains the perihelion of Mercury, that is, its deviation from what Newtonian mechanics predicted, and in addition explains the results of the famous experiment conducted by Eddington in 1919 that measured the bending of light in the sun's gravitational field. General relativity was rela-

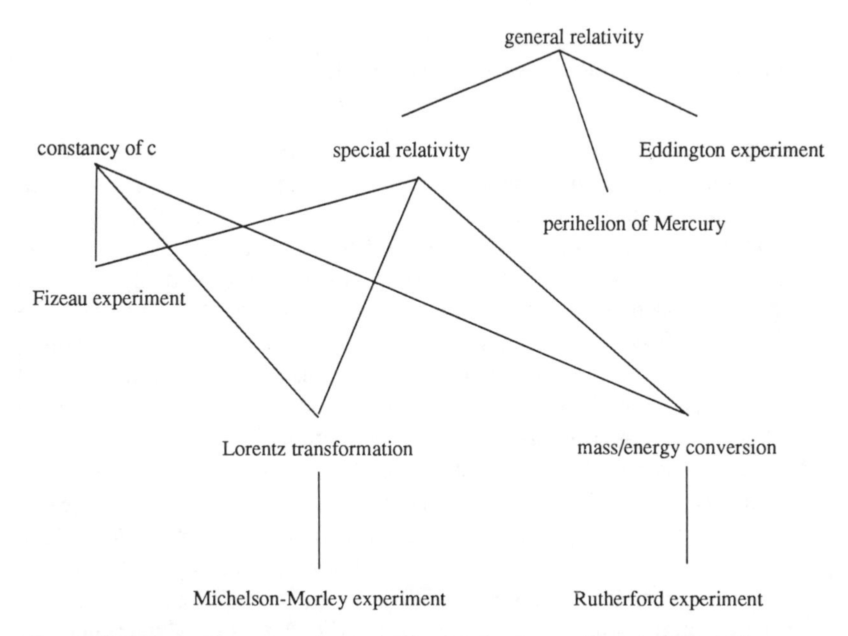

Figure 8.7. The explanatory coherence of relativity theory. Straight lines indicate explanatory relations.

tively lacking in experimental support for some decades after Einstein proposed it, but now explains a wider range of phenomena such as time delays in radar signals close to the sun (Will 1986).

According to some philosophers of science (Popper 1959; Lakatos 1970), prediction of new observations provides more support for a theory than explanation of facts already known. Then the Eddington experiment would contribute more to the acceptability of general relativity than its explanation of the perihelion of Mercury. This was clearly not Einstein's view of the matter, and Brush's detailed historical investigations show that it was not the view of most of his contemporaries either:

> Most of the published comments by physicists during the first 2 or 3 years after the 1919 eclipse observation indicated that light bending and Mercury's orbit counted equally strongly in favor of general relativity. If light bending was more important that was not because it had been forecast in advance, but because the data themselves were more definitive. . . . It later became clear to the experts that the Mercury orbit was stronger evidence for general relativity than light bending. In part this was because the observational data were more accurate—it was very difficult to make good eclipse measurements, even with modern technology—and in part because the Mercury orbit calculation depended on a "deeper" part of

> the theory itself. The fact that light bending was a forecast [prediction] whereas the Mercury orbit was not seems to count for little or nothing in these judgments. (Brush 1989, 1126)

Brush points out that Einstein's explanation of Mercury's orbit was all the more impressive because the discrepancy with Newtonian theory had been known for decades: other theorists had been unable to give a Newtonian explanation without making implausible ad hoc assumptions.

Figure 8.7 shows that relativity theory coheres well internally and with important evidence, but what is its relation to the previous hypotheses? There is no simple answer to this question, since we have to describe the relation of both special and general relativity to the theories of Newton, Maxwell, and Lorentz. Of these, Maxwell's equations come out unscathed, although thc background assumption that they describe motions in an aether is rejected. Similarly, many of Newton's background assumptions about absolute space and time are rejected, but his key principles can be retained for special cases that do not involve huge velocities or masses. Let us consider Newton's three laws and principle of gravitation. The first law of motion says that a body at rest tends to remain at rest unless acted on by an outside force, and similarly for bodies in motion. This survives intact within the Einsteinian framework, except that "at rest" can no longer be understood in any absolute sense. Similarly, the third law, that every action has an equal and opposite reaction, is not challenged by relativity theory. In contrast, the second law, relating acceleration, force, and mass by the equation $F = ma$ requires significant modification. Taking acceleration as rate of change in velocity v, Feynman, Leighton, and Sands (1963–1964, vol. 2, 18-2) write it as:

$$F = \frac{d}{dt}(p), \textit{ where } p = \frac{mv}{\sqrt{1 - v^2/c^2}}$$

Here the rate of change is modified by the familiar factor taking into account the velocity and the speed of light. Unless velocity is very high, $1 - v^2/c^2$ reduces to one so the equation reduces to Newton's. Similarly, Feynman et al. (1963, vol. 2, 42-6) express the curvature of space in general relativity by an equation relating the excess radius of a sphere produced by the curvature to $GM/3c^2$, where G is the gravitational constant and M is the mass of the matter inside the sphere. Clearly, if M is small, the curvature of space will be negligible and in accord with Newtonian mechanics.

By virtue of the mathematical relations described in the paragraph before last, the special and general theories of relativity are able to explain why Newtonian mechanics holds for the familiar cases in which velocities and masses are not enormous. Thus the Newtonian principles are coherent with the Einsteinian ones, rather than contradicting them. Incoherences in the whole system involving both pre-Einsteinian and Einsteinian hypotheses do, however, arise, since relativity rejects the existence of absolute frames of reference such as that provided by the aether. In addition, there were incoherences within the Newton-Maxwell theory, as we saw in the discussion of the asymmetries that Einstein cited as part of what spurred him to produce the theory of relativity. The relation between relativity theory and the views that preceded it is therefore complex, involving both incorporation of equations as special cases and rejection of assumptions central to Newtonian mechanics.

Because the applicability of Newtonian hypotheses to everyday objects follows from relativity theory, we can say that Einstein's theory *sublates* Newton's rather than supplants it (see section 5.1 for these distinctions). Sublation involves both incorporation and replacement. Because of relativistic rejection of some Newtonian concepts and severe modifications of others such as mass, it would be wrong to say that relativity theory simply incorporates Newtonian mechanics. But Einstein's revolution clearly preserves far more of the theory it replaced than did the other revolutionary theories so far discussed. The chemical, Darwinian, geological, Copernican, and Newtonian revolutions do not retain the superseded theories as holding for a limited range of phenomena, as relativity theory does with respect to Newtonian mechanics. But the final revolution in physics I want to discuss, in which quantum theory challenges Newtonian mechanics, is like Einstein's revolution in that Newtonian mechanics is sublated rather than supplanted.

8.4 QUANTUM THEORY

8.4.1 The Development of Quantum Theory

Twentieth-century physics rests on two cornerstones, general relativity theory and quantum theory. Quantum theory also developed early in this century, primarily between 1900 and 1927. Again I will not attempt a full historical description (Jammer 1966; Guillemin 1968; Gamov 1966; Kuhn 1978), but will sketch the key developments relevant to questions of conceptual change.

The scientific background to the development of quantum theory, as for relativity theory, consists of Newtonian mechanics and Maxwell's electrodynamics, both well established at the end of the nineteenth century. Unlike relativity theory, however, the origins of quantum theory are more explanation-driven than coherence-driven, although later developments appear to be coherence-driven. Another difference is that whereas relativity theory was largely the creation of Einstein, quantum theory derives from the efforts of a host of physicists, especially Max Planck, Einstein, Louis de Broglie, Niels Bohr, and Werner Heisenberg. Like the geological revolution, but unlike the other scientific revolutions we have considered, the quantum revolution was inspired collectively.

The origins of quantum theory lie in the problem of blackbody radiation. Glowing solids and liquids emit light with wavelengths depending on the temperature of the body. An ideal radiating body is one capable of emitting light of every wavelength; since such an emitter is also an ideal absorber, it is called a *black body*, and what it produces is called blackbody radiation. In 1900, Planck developed a formula for E_λ, the radiation energy emitted per second for wavelength λ:

$$E_\lambda = \frac{C_1\lambda^{-5}}{\exp(C_2/\lambda T) - 1}$$

where C_1 and C_2 are constants and *exp* means *e* to the exponent that follows (*e* is the base of the natural logarithms, about 2.718). Planck did not derive this formula from data, but did mathematical transformations on another formula that had been derived from data, so the discovery of this law is adequately characterized as data-driven (Langley et al. 1987, 49–53).

Planck was not content merely to find the formula that fit the data, however, but set out to find physical principles from which the formula could be derived. According to Holton and Brush (1985, 432), Planck was convinced that radiation must be explained by the action of submicroscopic electric oscillators. His radical new hypothesis was that the energy sent out by an oscillator of frequency *f* can only be an integral multiple of the quantity *hf*, where *h* is a universal constant, now called Planck's constant, approximately equal to 6.625×10^{-27} erg-seconds. This hypothesis was the birth of the idea that radiation is quantized rather than continuous as the wave theory of light and Maxwell's electrodynamics would suggest. Using this hypothesis and mathematical techniques borrowed from Boltzmann, Planck was able to explain the above equation. The formation of the hypothesis involving discrete energy levels is accurately described as explanation-driven, so we can call it abductive, although the mathematical methods underlying its formation are much more complicated than ones any current abduction programs can perform.

Blackbody radiation concerns light *emitted* by bodies, but Einstein in 1905 used Planck's constant to explain the photoelectric effect, which concerns the *absorption* of light. In 1897, J. J. Thomson had postulated the existence of the *electron* to explain numerous electric phenomena, and it was known that various materials emit electrons when exposed to light. Emission, however, does not conform to what one would expect from Maxwell's laws, since even feeble light over a minimum frequency can produce an electric current. In Maxwell's system one would expect a delay for the energy of a light wave, uniformly distributed over the whole wave front, to accumulate sufficiently to move the electron. Einstein postulated that the transformation of light energy into the kinetic energy of electrons was also limited by Planck's quantity *hf*. A quantum of light, later called a *photon*, may or may not give an atom enough energy to loosen an electron. Like the formation of Planck's hypothesis, Einstein's discovery was explanation-driven but highly mathematical. Treating light as quantized flew in the face of the established wave theory, requiring the revolutionary conceptual changes to be described in the next section.

Discovery of the electron as a component of atoms had shattered the ancient view of atoms as indivisible. Using a planetary analogy that had been previously proposed, Ernest Rutherford argued from experiments involving particle scattering that atoms consisted of a tiny positive nucleus orbited by electrons. In 1913, Niels Bohr showed how to combine the nuclear idea of the atom with quantum theory. He postulated that an electron can only radiate a light quantum or absorb energy while moving from one stable orbit to another. The change in energy is equal to hv, where h is Planck's constant and v is the velocity of an emitted photon. Electrons are thus restricted to a limited range of energy levels, which explains why light is absorbed and emitted at discrete intervals, as shown for example by lines in the spectrum for light emitted from hydrogen.

Early on, quantum theory developed largely within the boundaries of classical Newtonian/Maxwellian physics. To use mathematical techniques from classical physics in the quantum domain, Bohr formulated a *correspondence principle* that maintained that classical laws are approximately true in the quantum domain. The early ideas developed by Planck, Einstein, and Bohr are usually referred to as the "old" quantum theory.

Much more radical views came forth in the 1920s. Einstein had postulated that light, assumed to consist of waves, had quantized, particle-like properties. In 1923, de Broglie made the shocking suggestion that particles such as electrons behave in part like waves. An electron of mass m and velocity v will have a wavelength λ given by:

$$\lambda = \frac{h}{mv.}$$

Since Planck's constant h is very small, only particles with very small mass such as electrons will have a wavelength of any consequence. Nevertheless, de Broglie's formula produced a startling alteration in the traditional division between particles and waves. From his own description, it appears that the postulation of the wave nature of particles by de Broglie was coherence-driven. By 1923, Einstein's hypothesis of light quanta had been used to explain many phenomena such as the Compton effect, so de Broglie was convinced that radiation was discontinuous, in conflict with the classical wave notions. "It was necessary to assume willy-nilly that the picture of waves and the picture of corpuscles had to be used one after the other for a complete description of the properties of radiation" (de Broglie 1953, 158). De Broglie audaciously attempted to increase the coherence of physics by suggesting that even particles have wave-like properties, thereby dissolving the division between waves and particles.

A year later, Schrödinger developed a general mathematical theory of atomic properties called *wave mechanics* that explained the motion of electrons within atoms, including spectral frequencies. Shortly afterwards, using a different notation, Heisenberg developed quantum (matrix) mechanics, which Schrödinger proved equivalent to his wave mechanics. The seminal developments of de Broglie, Schrödinger, and Heisenberg were certainly not data-driven, nor were they provoked by the need to explain any particular phenomena. Rather, they were largely coherence-driven in that they served to tie together quantum ideas. Later, however, quantum theory would receive an enormous amount of evidential support.

Additional developments in the mid-1920s occurred that are often taken to be essential to the new quantum theory: Heisenberg's uncertainty principle and Bohr's principle of complementarity. Neither of these is really a hypothesis of quantum theory: the uncertainty principle was derived mathematically by Heisenberg from his quantum mechanics, while Bohr's complementarity claim is an interpretation of quantum theory. In 1927, Heisenberg showed that according to quantum theory there are limitations on how accurately certain pairs of physical properties can be measured. One such pair is the position and momentum of an electron. If Δx is the error of measurement of an electron's position, and Δp is the error of the measurement of its momentum, then the product of these errors can never be reduced below an amount determined by Planck's constant h:

$$\Delta x \cdot \Delta p \geq \frac{h}{2}\pi$$

The more accurately you measure the position of an electron, the less accurately you can measure its momentum. Since linear momentum is the product of mass times velocity, this is equivalent to saying that you cannot simultaneously give the location and velocity of a particle. A similar limitation applies to measuring the energy of a particle and the time interval during which it has that energy. These limitations undercut the determinism that was implicit in Newtonian mechanics, since they place a bound on the extent to which the motion of very small particles can be predicted.

According to Bohr's principle of complementarity, proposed in 1927, the wave and particle descriptions of matter are equally good, even though they might seem incompatible to a Newtonian. He attempted to elevate the use of incompatible sets of descriptions to a general principle of scientific practice, but complementarity has remained obscure. Jammer (1966) finds its origins in the idealist philosophy of Bohr's friend Harald Höffding. According to idealism, the world is essentially mind-dependent. Some physicists have given an idealist interpretation of Heisenberg's uncertainty principle, saying that the impossibility of measuring the exact location and momentum of a particle means that it does not really have these properties *until* we choose to measure one of them.

Claims about the mind-dependence of the world are not new. Kant (1965) claimed to be enacting a "Copernican revolution" through his doctrine that the mind makes a priori contributions to the understanding of experience, but his was if anything a "Ptolemaic revolution," putting humans back at the center of the knowable world. The Copenhagen interpretation of quantum theory is similarly Ptolemaic, in contrast to the trend from Copernicus through Darwin and Einstein to remove humanity from the center of existence. There is a sharp contrast between Einstein's relativity principle, which says that the laws of physics are *not* relative to people's frame of reference, and the Copenhagen interpretation's emphasis on the dependence of reality on observation. But there are other interpretations of quantum theory besides the Copenhagen version, which is not required for the explanatory application of the theory.

8.4.2 Conceptual Change in Quantum Theory

As with relativity theory, it is possible to minimize the extent of the difference between quantum theory and classical physics by looking only at simple mathematical relations. Consider the last two equations. If the value of Planck's constant h were 0, then the significance of both de Broglie's wavelength equation and Heisenberg's uncertainty principles would disappear. Particles would have 0 wavelength and there would be no limit to the extent to which the error in measuring both position and momentum could be reduced. Moreover, since h is so small, for objects of everyday experience it has essentially no relevance. Just as classical physics survives relativity theory if one avoids very fast or very large objects, so classical physics survives quantum theory if one avoids very small objects. It would be a mistake, therefore, to say that quantum theory replaces classical physics.

But it would also be a mistake to say that quantum theory simply extends classical physics, since there are definite incompatibilities. We find associated with quantum theory such radical pronouncements as that light is a particle and that particles have wavelength, both flying in the face of the nineteenth-century framework of classical physics. As with relativity theory, the relation between quantum theory and classical physics is best described as *sublation*, as both a rejection and a preservation. The preservation consists in explaining why classical physics works so well in its appropriate sphere, but the cancellation consists in proposing new principles organized into a different and incompatible conceptual system.

The classical view distinguished between two distinct kinds of entities: matter formed of atoms, and radiation of different kinds differing only in wavelength. Starting with Thomson's discovery of the electron and moving through all the developments of quantum theory, the simple kind-relations and natural part-relations of classical physics were radically altered. Consider first the part-relations and the concept of an atom. The concept of the atom as the fundamental constituent of matter originated with the ancient Greek philosophers Leucippus and Democritus; the word "atom" derives from the Greek word meaning indivisible. The concept of the atom was not part of the eighteenth-century chemistry of Lavoisier or the phlogiston theories, but became influential early in the following century through the work of John Dalton. Through the work of Thomson, Rutherford, and Bohr, the concept of an atom was changed from assuming an indivisible whole to a nonfundamental entity consisting of parts: a nucleus with orbiting electrons. Bohr's introduction of quantum levels for the electrons took the concept of an electron a substantial step away from Rutherford's visualizable planetary system, and subsequent developments in quantum theory have proceeded without any simple physical picture. The components of the atom, including the nucleus with protons and neutrons, and electrons operating at various energy levels, do not bear to the atom the part-relation familiar from everyday life. Advanced physics students are warned not to attempt to visualize the atom as a core nucleus with orbiting electrons. The early developers of the theory of atomic structure, Thomson and Rutherford, produced a less radical kind of conceptual change, part-whole decomposition. But later mathematical treatments of atoms appear to alter standard notions of part-relations, producing the conceptual change I have called tree switching.

Part-relations are also disturbed by Heisenberg's uncertainty principle, which affects our basic concepts of space and time. De Broglie (1953, 178) wrote that with wave mechanics we can no longer avail ourselves without precautions of the notions of a position, velocity, and trajectory of a particle. We saw that relativity theory required amalgamation of Newtonian ideas about space and time into space-time, but even in space-time objects have a definite location along the four dimensions. Heisenberg's uncertainty principle can be interpreted as saying that a particle has no definite position until it is observed, or less idealistically as saying that its position is inherently statistical. Either way we abandon the classical commonsense view of particles having a definite place in space and time. Similarly, the notions of part and whole become more indeterminate than they were classically, since spatial and temporal parts are no longer absolutely determinable.

Einstein and other physicists rejected the indeterminism that Heisenberg and others saw as essential to quantum theory. There are reasons for this rejection that go beyond mere questions of part-whole relations. Einstein's metaphysical view of the universe as a whole prevented him from accepting the consequences of quantum theory concerning uncertainty. Most physicists, however, were more flexible and it did not take long for the fundamental notions of quantum theory to be established.

Another major conceptual change associated with quantum theory involves the concepts of wave and particle. The wave-particle dichotomy had been an issue ever since the Greek Stoic Chrysippus first developed the wave theory of sound around the third century B.C. In the seventeenth century, Christian Huygens argued, by analogy with sound, for a wave theory of light. But Newton's particle theory of light dominated until Thomas Young and Augustin Fresnel established a more sophisticated wave theory of light in the first half of the nineteenth century. Maxwell unified the wave theory of light with Faraday's field theory of magnetism, producing the concept of electromagnetism. Within Newtonian mechanics, however, objects such as the planets were treated essentially as point masses, corpuscles with nothing in common with phenomena of light or magnetism. Hence Einstein's hypothesis that light is quantized and de Broglie's suggestion that particles have wavelengths represented major alterations in the kind-hierarchy of classical physics. Both Bohr's complementarity principle that waves and particles are incompatible but equally admissible ways of viewing phenomena, and the less idealistic view that particles such as electrons are neither waves nor particles but require a new category, imply that the neat classical taxonomy of entities into waves and particles has foundered. A wave can be understood, not as a concrete thing like water waves, but as anything that satisfies a wave equation. Electrons are particles obeying Schrödinger's equation, whose solution is a wave. Students still have difficulty trying to conceive of something as both a wave and a particle, but the large amount of experimental evidence for quantum theory suggests strongly that the old sharp wave/particle distinction is gone for good.

8.4.3 The Explanatory Coherence of Quantum Theory

Despite the major conceptual changes that it required, quantum theory gained general acceptance. By 1920, the quantum ideas of Planck and Einstein had generally been accepted (Holton and Brush 1985, 491), and the new quantum theory came to dominate physics in the 1930s. It explained numerous phenomena that classical Newtonian physics could not: the distribution of energy in the spectrum of blackbody radiation, the photoelectric effect, the stability of atoms, the discrete spectrum of hydrogen, the diffraction of electrons, and so on. Quantum electrodynamics, developed in the 1940s and 1950s, succeeded in explaining cosmic ray showers, the anomalous magnetic moment of the electron, superconductivity in metals, the superfluidity of helium, and the quasi-stable configuration of an electron and positron (Bynum et al. 1981, 354). It would be informative to trace out how the different principles of quantum theory, including quantum electrodynamics, are used in the explanations of the many important phenomena they explain.

Quantum theory should be seen as sublating Newtonian physics rather than supplanting it. Numerous changes in kind-relations and part-relations were required in the development of quantum theory, but it did not challenge the applicability of Newtonian mechanics to everyday objects with mass much greater than Planck's constant. In contrast to phlogiston theory, creationist biology, earth contraction theory, and Aristotelian/Ptolemaic physics, Newtonian physics, in its restricted and modified form, remains part of current scientific doctrine.

8.5 FORCES, QUARKS, AND SUPERSTRINGS

Relativity theory and quantum theory were established more than fifty years ago, and obviously much has happened in physics since. Several developments are particularly interesting for conceptual change.

The Bohr atom consists essentially of a positive nucleus and electrons. In 1932 James Chadwick discovered the uncharged particle the neutron, dividing the nucleus into positive protons and neutrons. In the same year, Paul Dirac's hypothesis of the positron, or positive electron, was confirmed experimentally. A proliferation of so-called fundamental particles followed: the neutrino, the muon, and dozens more. To bring order to the mounting chaos, Murray Gell-Mann and George Zweig proposed that many of these particles were made of still smaller particles called quarks. On the current view particles divide into two kinds: hadrons, such as the proton and neutron, which consist of quarks; and leptons, such as electrons, which are indivisible (Davies 1989). All matter is made of quarks and leptons. The kind-hierarchy of kinds of particles has expanded dramatically in the past few decades, and a new layer in the part-hierarchy has added quarks as parts of particles previously taken to be fundamental. The concept of particle has thus acquired both new kinds (hadron, lepton, and the more than a hundred particles classified under them) and new parts, the quarks. Hadrons are divided into mesons, which have only one quark, and baryons that have three. However, a proliferation of varieties or "flavors" of quarks has made some theorists suspect that they are not fundamental either.

Whereas classical physics postulated two kinds of force, gravitational and electromagnetic, current theory postulates two additional kinds operating at the subatomic level. The "strong" force binds hadrons such as neutrons and protons together, while the "weak" force is involved in radioactive decay. We thus have a differentiation, in the sense of Chapter 3, of the concept of force. Theorists, however, have been laboring to unify the four forces in a so-called Grand Unified Theory. So far, the electromagnetic and weak forces have been shown to be fundamentally the same, with coalescence producing the concept of the *electroweak* force. But unification with the gravitational and strong forces has eluded scientists' attempts. In the history of the concept of force, we see a series of differentiations and coalescences, and a grand unified theory would provide the ultimate coalescence.

I described general relativity and quantum theory as the two cornerstones of twentieth-century physics, but theorists have been frustrated in their attempts to integrate the two. According to Edward Witten, "the basic problem in modern physics is that these two pillars are incompatible. If you try to combine gravity with quantum mechanics, you find that you get nonsense from a mathematical point of view. You write down formulae which ought to be quantum gravitational formulae and you get all kinds of infinities" (Davies and Brown 1988, 90). He advocates a new approach, *string theory*, to try to overcome the inconsistency between gravity theory and quantum mechanics. There are no special phenomena that string theory is supposed to explain, so its development is clearly coherence-driven rather than data- or explanation-driven. Witten compares attempts to reconcile relativity and quantum theories to the attempt by Einstein to reconcile Newton and Maxwell. Because string theory is in the early stages of development, we have no way of knowing if it will indeed provide a new synthesis that will sublate relativity and quantum theories. But if it succeeds, string theory will not simply be an extension of those theories, since reconciling them will require substantial conceptual change. According to the new views, an electron is not a particle, but a little vibrating string operating in extra dimensions that make sense of its gravitational field. Conceiving of electrons as strings and space-time as having ten or more dimensions means that acceptance of string theory would bring with it major changes in kind-hierarchies and part-hierarchies. Some theorists have complained that superstring theory as yet makes no new predictions testable by experiment, but bringing coherence to quantum and relativity theories is a laudable goal in itself. Increases in explanatory coherence from new evidence often take a while to develop, as with general relativity. Physical experiments are becoming more and more expensive to perform as ever higher energy levels are required, but there is no reason to believe that physics has seen its last conceptual revolution.

8.6 SUMMARY

The development of physics from Aristotle to relativity and quantum theories has required substantial changes in concepts such as force, gravity, mass, planet, wave, and particle. Copernicus supplanted Ptolemaic astronomy with a broader and simpler explanatory system. Newton completed the demise of Aristotelian physics by developing a highly unified theory that explained both terrestrial and celestial phenomena, and he supplanted Descartes's vortex theory of gravity. In the twentieth century, the relativity and quantum theories showed the inapplicability of Newtonian theory to objects that are very massive, small, or fast-moving. The revolutions produced by these theories were more cumulative than other scientific revolutions, but they still involved considerable conceptual change and rejection of previously held views. In all four revolutions in physics, the replacing theory had greater explanatory coherence than the one it replaced.

CHAPTER 9

Revolutions in Psychology?

THE previous chapters have analyzed seven revolutions in the natural sciences, covering physics, chemistry, biology, and geology. Each revolution has naturally been described in terms of changes in conceptual hierarchies produced by the acceptance of a new theory with greater explanatory coherence than the old. New theories brought with them major conceptual reorganizations. The question naturally arises whether the same sort of analysis applies to developments in the social sciences: psychology, economics, politics, sociology, and anthropology. Examination of the recent history of psychology suggests that conceptual change in the natural sciences differs from that in the social sciences, which have not yet developed to the point where they have a coherent unifying theory to be overthrown.

The main topics of this chapter are the two major developments in twentieth-century psychology, behaviorism and cognitivism. Writers often refer to the "cognitive revolution," and histories sometimes also refer to the "behaviorist revolution" that preceded it. Although both developments involved conceptual change, they differed from revolutions in the natural sciences. Behaviorism and cognitivism are better described as competing *approaches* than as competing explanatory theories. Their adoption was more the result of methodological considerations than evaluations of explanatory coherence.

9.1 APPROACHES, FRAMEWORKS, AND THEORIES

ECHO analyses in previous chapters have shown how theories, understood as coherent collections of hypotheses, serve to explain a broad range of empirical generalizations and facts. Psychology has not yet achieved a unifying theory comparable to Newton's mechanics, Lavoisier's oxygen theory, or Darwin's theory of evolution by natural selection. Despite the undeniable advances in the understanding of human thinking made in this century, we still lack a unified theory that explains a wide range of phenomena. Appreciation of this fact is crucial for understanding the conceptual development of modern psychology, so a short diversion is necessary to make clear the differences between theories, empirical generalizations, frameworks, and approaches.

An approach is a general collection of experimental methods and explanatory styles. Approaches are thus much more general and diffuse than theories that explain identifiable sets of facts. Psychology has seen several major approaches in its brief history since the first psychology laboratory was founded by Wilhelm Wundt in 1879. Wundt and his American disciples such as Edward Titchener are called *introspectionists*, since their primary experimental method consisted in having people report their mental experiences. When J. B. Watson proposed *behaviorism* in 1913, he explicitly rejected this method, arguing that psychology should concern the objective observation of behavior, not the subjective introspection of experience. In the mid-1950s, *cognitivism* relegitimated the discussion of mental representations and processes and the performing of experiments that addressed questions about such nonobservable constructs.

Within a single approach, there can be various frameworks for investigating the phenomena deemed relevant by the approach. B. F. Skinner, for example, developed a framework for psychological investigation that relied more on his notion of operant conditioning than on Pavlovian conditioning, which was Watson's principal investigative technique. (These concepts are discussed below in section 9.2.1.) Within cognitivism today, there are a variety of frameworks, including the interpretation of the results of psychological experiments in terms of mental representations involving rules. In section 9.3.4 I shall characterize connectionism as another framework within the cognitivist approach. Although frameworks are more specific than approaches, they are still much vaguer than theories that have identifiable hypotheses and ranges of explanation. As I use the terms, a theory can be part of a framework that assumes an approach.

We also need to distinguish between theories and *empirical generalizations*. Some psychologists use the term "theory" very loosely, encompassing any general belief that ordinary people might have. My usage is more specific and more compatible with usage by natural scientists and philosophers of science. In physics, for example, we speak of Kepler's *laws*, but Newton's theory. Kepler's laws are empirical generalizations that summarize observations of planetary motion. In contrast, Newton's principles go well beyond observables with theoretical hypotheses such as that there is a force of gravity between any two objects. Similarly, Darwin's postulation of natural selection involved a nonobserved process that served to explain a wide range of empirical generalizations such as those concerning the fossil record. Figure 9.1 concisely depicts the kind-hierarchy of representations I have in mind. Theories are not simply sets of propositions, but structures encompassing conceptual relations and past problem-solving successes in addition to hypotheses (Thagard 1988; section 4.4 above). Hence theories have both propositional and nonpropositional representations as parts. By "proposition" I mean a mental representation similar to a sentence (section 2.3). Some sentences describe particular facts, such as that Lassie the dog has fur. Generalizations are universal statements, for example that all dogs have fur. In AI systems, gener-

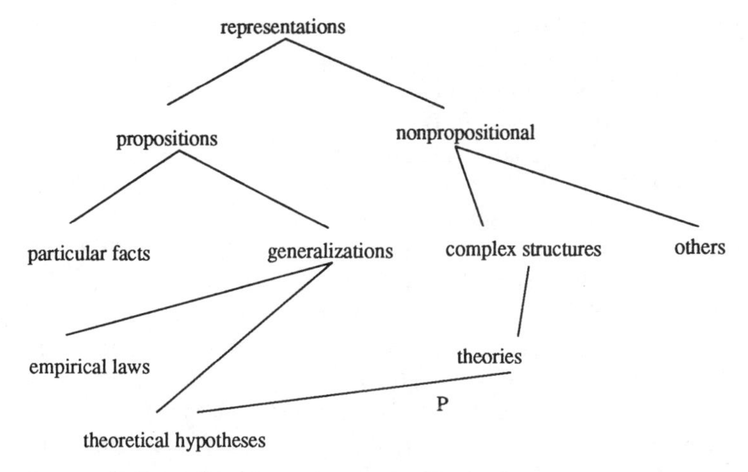

Figure 9.1. Taxonomy of representations. Straight lines indicate kind-relations, except for the part-relation labeled "P."

alizations are usually represented by rules (section 2.5). Some generalizations do not merely summarize observations but instead use theoretical concepts to provide explanations. Not shown in Figure 9.1 are other kinds of nonpropositional representations, such as pictures and the distributed representations discussed in section 9.3.4.

Psychology has witnessed few attempts at general theories such as Clark Hull's learning theory discussed below. Cognitive psychology has local theories that postulate nonobservable representations and information-processing mechanisms to explain experimental results; the results explained are more commonly called "effects" than generalizations. But no precise theory that serves to unify and explain a wide range of experimental results has emerged. In looking at behaviorism and cognitivism, we will see important conceptual changes, even in the absence of unifying explanatory theories. Of particular interest will be the relation between behaviorism and cognitivism, and the relation each has to the commonsense psychology of ordinary people.

9.2 BEHAVIORISM

9.2.1 The Development of Behaviorism[1]

An early textbook by a prominent American psychologist gives the flavor of the approach to psychological research against which the behaviorists rebelled. "Psychology is commonly defined as the science of consciousness. . . .

[1] A fuller historical account would discuss many matters omitted here. Historians of psychology distinguish the "structuralist" approach of Wundt and Titchener, whose goal was to discover

Mental facts, or facts of consciousness, constitute the field of psychology. . . . The fundamental psychological method is introspection. . . . As a psychological method it consists simply in the direct examination of one's own mental processes" (Angell 1908, 1–5). The founder of behaviorism, J. B. Watson, did his early research with animals, particularly rats running mazes. He inverted the introspectionist's transfer of notions of consciousness to animals by denying that the topic of consciousness is relevant even to the psychology of humans. In his original manifesto, "Psychology as the Behaviorist Views It," he advocated the "elimination of states of consciousness as proper objects of investigation" (Watson 1913, 177). The goal of psychology is merely the "ascertaining of such data and laws that, given the stimulus, psychology can predict what the response will be; or, on the other hand, given the response, it can predict the nature of the effective stimulus" (Watson 1917, 337). Abandonment of the method of introspection also meant abandoning concern with consciousness, sensation, perception, attention, will, image, and the like (Watson 1929, xii). Watson did not completely abandon other concepts such as thinking and memory, but redefined them so as not to refer to the inner life of the person. Thinking, for example, was redefined as "subvocal talking" and memory became the retention of bodily habits.

According to Watson, psychology was concerned only with empirical laws, not with explanatory theories. He urged: "Let us limit ourselves to things that can be observed, and formulate laws concerning only those things" (Watson 1959, 6). Psychology should study learned acts called *habits*; a principal means by which habits are acquired is the *conditioned reflex* described by the Russian psychologist Ivan Pavlov. Here one stimulus comes to stand for another that had originally initiated a reflex response, as when a bell comes to stand for food and evokes salivation. Behaviorist psychologists, however, were not limited to studying conditioned reflexes, since they could also observe verbal reports and the results of psychometric tests. For Watson, the study of verbal reports differed from the introspection of prebehaviorists, since such reports were simply understood as behavior. Verbal reports were just data to be analyzed, not direct descriptions of anything mental.

Watson's research career ended in the 1920s, after he was fired from Johns Hopkins for an affair with a graduate student. Even with him on the sidelines, many psychologists moved more and more away from introspection and in-

---

the structure of mental experience, from the "functionalist" approach that originated with William James and John Dewey. The latter emphasized the biological functions of the mind in controlling behavior, and thus can be viewed as a precursor to behaviorism, although it did not eschew introspection. James advocated using introspection to describe mental phenomena, but also attempted to give evolutionary explanations of these phenomena in terms of their functions. My description of the rise and dominance of behaviorism primarily describes the American scene. In Europe, where psychologists such as Bartlett, Piaget, and Wertheimer were influential, behaviorism did not reign as it did in the United States.

creasingly toward the "objective" study of behavior. Research on animals, which was peripheral when Watson began his career, moved to the center of experimental psychology. By the 1930s, behaviorism was established as the dominant approach in American psychology. Clark Hull was the most influential behaviorist of the 1930s and 1940s, while B. F. Skinner rose to prominence later.

Whereas Watson recommended merely forming laws about stimuli and responses, Clark Hull modeled his theoretical efforts after Newton. He viewed a theory as "a systematic deductive derivation of the secondary principles of observable phenomena from a relatively small number of primary principles or postulates" (Hull 1943, 2). Hull allowed the postulation of *intervening variables* between observable stimuli and responses, offering the schema:

$$S \rightarrow s \rightarrow r \rightarrow R$$

where the small "s" and "r" are neural processes relating the stimulus S and the response R. Laws relating S and R are to be expressed in terms of *habit strength*. Increases in habit strength are the result of the *reinforcement* of the response under appropriate conditions. Although Hull was methodologically much more sophisticated than Watson and saw the need for psychology to have a general unifying theory, the theory he proposed was unsuccessful, even by behaviorist standards. A host of experimenters failed to get the experimental results predicted by the theory, which became increasingly complex as it had to be altered in an attempt to deal with anomalies. Hull's attempt to become the Newton of psychology was a failure.

Skinner's approach was a throwback to Watson's. Eschewing Hull's theoretical apparatus, he returned to the project of finding empirical generalizations that relate stimuli and responses. With Watsonian vehemence, he contended that inner states have no place in psychology. (In fact, Skinner's approach was even more extreme than Watson's, since it did not allow the discussion of thought in terms of subvocal speech.) At the center of psychology was the study of learning, understood as the "reassortment of responses in a complex situation" (Skinner 1953, 65). Skinner distinguished *operant conditioning* from Pavlovian conditioning in which stimuli are paired. In operant conditioning, the experimenter waits for an animal spontaneously to show a desired response that is then reinforced. For Skinner even more than Watson, thinking, reasoning, and emotion are fictional causes that need not be postulated in the scientific study of behavior. As with Hull's ideas, Skinner's views led to much experimentation, most of it with animals. Not all psychologists who considered themselves behaviorists were as theoretically abstemious as Skinner and Watson. Edward Tolman (1932) saw no problem with discussing animal behavior in terms of purpose and cognition, so long as these were reinterpreted in behaviorist terms.

9.2.2 Behaviorism and Conceptual Change

People are all psychologists in their ordinary lives, in that everyone makes generalizations about and tries to explain the actions of other people. Embedded in our everyday belief system is a rich array of generalizations and explanatory assumptions. For example, I may explain a friend's agitation using the knowledge that she is writing her dissertation and the inference that she is anxious about finishing in time. I attribute to her the mental state of anxiety which is not directly observable. Such attributions are theoretical in the same way that postulation of electrons is theoretical, although physics is much easier than commonsense psychology to boil down to explicit explanatory hypotheses. It would be a daunting task to write down even some of the generalizations that people use in explaining the behavior of themselves and others. Social psychology has barely scratched the surface in investigating the beliefs that we use in accounting for people's actions, although there is a growing consensus that the domain is one of rich causal knowledge. Wellman (1990) provides a useful outline of commonsense psychology.

Much of prebehaviorist psychology is a systematic extension of commonsense psychology. Although it had a physiological side and introduced notions like stimulus and response that are not part of commonsense psychology, most of introspectionist psychology was compatible with what ordinary people believe about the mind. Books like Angell (1908) and James (1961) had chapters on such topics as perception, imagination, memory, feeling, and reasoning, with insights that surpassed those of commonsense psychology but certainly did not supersede them. Behaviorism, in contrast, involved a massive rejection of commonsense psychology along with introspectionism. It of course also rejected theoretical constructs that went beyond commonsense psychology, such as the Freudian unconscious.

Behaviorism is more noteworthy for conceptual deletions than for additions. The basic concepts of behaviorism—stimulus, response, conditioning, reinforcement—were already part of psychology through animal research by Pavlov, Thorndike, and others. What was radical about behaviorism was the claim that the range of these concepts could be extended to cover humans and could obviate the concepts of commonsense psychology. The polemics of Watson and Skinner advocate jettisoning such staples of commonsense psychology as consciousness, will, imagination, and so on. Eliminated with them would be the vast parcel of explanatory generalizations concerned with mental states. Never in the history of science has conceptual extinction on such a vast scale been attempted.

Behaviorists did not, however, reject all the concepts of commonsense and introspectionist psychology, since some could be reinterpreted in behaviorist terms. This reinterpretation often involved branch jumping, in which mental states and processes are recategorized as kinds of observable behavior. For Watson, thinking ceased to be a kind of mental activity, and instead was classified as a physical response. Language became a kind of verbal behavior instead of a kind of knowledge and mental processing. Learning, which dictionaries define in mental terms as an increase in knowledge or understanding, became a reassortment of responses. Thus the mental concepts of introspectionism and commonsense psychology that were not simply deleted underwent reinterpretation and reclassification within the restricted behaviorist conceptual framework of stimuli and responses.

Whereas behaviorism has a limited kind-hierarchy, classifying concepts as responses, habits, and so on, it rejects a concept of mind or mental process that has parts. A Freudian organization of mind into conscious and unconscious, or Id, Ego, and Superego, was unthinkable. Hence along with the large catalogue of mental activities that are part of commonsense psychology, behaviorism rejected the organization of these into part-hierarchies, eliminating an entire potential tree of concepts.

9.2.3 Behaviorism and Explanatory Coherence

Behaviorism might have been true. The behavior of humans and other animals might have been subject to simple principles statable in terms of stimuli, responses, and reinforcement schedules. If the conceptual system advocated by Watson and extended by later behaviorists had been adequate to explain all human behavior, it would have been one of the greatest scientific successes of all time, parsimoniously covering an enormous range of actions. But as the collapse of Hull's theoretical program showed, the behaviorist conceptual system was inadequate even for animal learning. Today, even Pavlovian conditioning is understood in terms of the expectations and internal representations of animals (Rescorla 1988). Rats, let alone people, turned out to be a lot more complex than the behaviorists expected.

Commonsense psychology has an enormous explanatory range, covering much of human behavior. For almost any human action, we can generate explanations:

He yelled at her because he was mad.
She stayed with him because she loved him.
He used insider information because he thought he could get away with it.

Although the coverage of commonsense psychology is great, it is sadly lacking in simplicity and unification, as these terms were explicated in Chapter 4. The number of explanatory principles seems to be almost as great as the number of behaviors. English contains thousands of terms describing traits, such as "brave," "generous," and so on. We use these terms gratuitously, saying that she fought fiercely because she was brave, or that he gave to the poor because he was generous. The behaviorists argued that such explanations are trivial redescriptions of the actions they explain. Commonsense psychology has thousands of rules that can contradictorily be fitted to almost any situation: haste makes waste; he who hesitates is lost. In contrast, behaviorism offered the possibility of simple explanations, but failed to give a unified description of the behavior of animals and humans because it explained so little.

In every revolution in the natural sciences, a new theory replaced or sublated a previous one because of greater explanatory coherence. It would be pointless, however, to try to do an ECHO analysis of behaviorism versus introspectionist or commonsense psychology. The prebehaviorist views did not constitute a unified theory with an identifiable set of hypotheses and evidential applications. Rather, they were (and are) a very large and amorphous set of propositions about human behavior and thinking. The preference of Watson and Skinner for low-level empirical laws rather than unified theories also tends to impede ECHO analysis. Hull's intentionally Newtonian approach could be analyzed in terms of explanatory coherence, but his crude use of pseudo-mathematical formulas and the failure of experiments suggested by the system makes a full analysis pointless. Freedman (1991) has used ECHO to analyze the debate between Hull and Tolman concerning latent learning.

So why did behaviorism come to dominate American experimental psychology? Behaviorism did not supplant introspectionism because of greater explanatory coherence, so why had so much of psychology turned behaviorist in the 1930s? Behaviorism, understood as an approach rather than as a theory, appealed for a variety of methodological reasons. The call to studying "objective" behavior rather than "subjective" experience made psychology sound more scientific. Moreover, the behaviorist research program gave psychologists lots to do. Academic industries grew up around the testing of Hull's learning theory and Skinner's operant conditioning. Scientists tend to move in directions where there seems to be interesting research to be done. Behaviorism is sometimes blamed on the logical positivists who dominated philosophy of science from the 1930s to the 1950s and shared the behaviorists' distaste for occult entities. But the behaviorists who were most directly influenced by the logical positivists, Hull and Tolman, were more methodologically sophisticated than Watson or Skinner. Behaviorism did not derive from logical positivism, although the two movements often found common cause (Smith 1986).

From the perspective of conceptual change alone, there *was* a behavioral revolution, since the conceptual system changed dramatically with the deletion of many mentalistic concepts and the reinterpretation of others in extremely revised kind-hierarchies. There is no point quibbling over the term "revolution," but it is crucial to keep in mind the difference between the ascension of behaviorism and the revolutions in the natural sciences. The rise of behaviorism consisted of the adoption of an approach for largely methodological reasons, not the acceptance of a theory for reasons of explanatory coherence. And so did the rise of cognitivism.

9.3 COGNITIVISM

9.3.1 The Development of Cognitivism

Behaviorism was not completely successful in banishing the mind from psychology. Tolman called himself a behaviorist, but had to withstand criticism from his peers for employing concepts such as *purpose* and *cognitive map* in his accounts of animal behavior. Behaviorism dominated the United States in mid-century, but abroad the study of cognition continued, for example in Jean Piaget's research on child development in Switzerland, Frederic Bartlett's work on memory in England, and the growth of Gestalt psychology in Germany. Even in the United States, social psychology remained a cognitive island in a behaviorist ocean, since leading figures such as Kurt Lewin and Fritz Heider felt free to discuss cognitive structures and processes. Nevertheless, in most psychological circles, behaviorism was the dominant approach into the 1950s.

The resurgence of cognitivism, and the rise of what is now called cognitive science, can be traced to three events in 1956 and 1957 (Gardner 1985; Baars 1986). In 1956, the psychologist George Miller published his influential paper on "The Magical Number Seven, Plus or Minus Two," which provoked new ways of looking at the structure of information. And Allan Newell and Herbert Simon announced their development of a computer program that proved theorems in logic. The following year, linguist Noam Chomsky's book *Syntactic Structures* gave rise to an approach to linguistics and the psychology of language very different from that of the behaviorists. I shall provide only enough of a sketch of the development of cognitivism to set the stage for discussion of the conceptual changes it involved.

Since the concept of *information* has been so important to cognitivism, let us begin with the communication-theoretic notion of information developed by Claude Shannon (Shannon and Weaver 1949). Information here is a property of a *signal*, as expressed by the equation:

$$I(s) = \log 1/p(s).$$

The information carried by a signal s is the logarithm (to the base 2) of the reciprocal of the probability of s, so that the signal is deemed to carry more information the less probable it is. The use of base 2 logarithms means that information can be measured in bits. This notion of information has virtually ceased to be employed by psychologists, but has recently been put to epistemological use by Fred Dretske (1981).

In psychology, information theory using the communication notion of information was influential in the early 1950s, but the prospects of applying the mathematical notion of information waned with Miller's (1956) seminal paper. Miller drew together numerous studies that showed that the channel capacity of human thinking seemed to be limited, with short-term memory, for example, restricted to around seven items. How, then, can people manage complex thinking with such restricted channels? Miller proposed that channel limitations are overcome because *chunks* of information, rather than bits, are what matter. By recoding stimuli into chunks, people overcome the informational bottleneck. Hence to understand thinking we need to look not just at bits of information but at the processes by which they are structured into chunks and at the nature of these encodings.

Around the same time, Newell, Shaw, and Simon (1958) were developing what they called an "information-processing" approach to thinking. Starting with the Logic Theorist, a program that proved theorems in formal logic, they developed ideas about how thinking could be viewed as directly analogous to newly developed computer processes. Their sense of "information" has only the vaguest relation to Shannon's mathematical notion. Modern cognitive psychology is often referred to as "information-processing" psychology, again with little relation to the mathematical notion of information. In a recent cognitive psychology textbook, John Anderson (1980, 13) characterized information as the "various mental objects operated on." Another textbook simply identifies information with "knowledge." Thus information-processing psychology, in keeping with Miller's notion of chunk but not with Shannon's mathematical notion, treats information as largely a matter of mental representation: computational structures in the minds of thinkers.

Behaviorism was challenged explicitly by Chomsky (1959) in his review of Skinner's 1957 book *Verbal Behavior*, and by Miller, Galanter, and Pribram's (1960) book *Plans and the Structure of Behavior*. Chomsky dissected Skinner's attempt to show how the learning and use of language could be understood in terms of operant conditioning. He argued that the concepts that Skinner introduced to describe the functional relations in verbal behavior were either inadequate to characterize human language or, when understood metaphorically, too vague to illuminate the phenomena. He argued instead that the knowledge of language acquired so quickly by a child should be understood in terms of a complex grammar, not a collection of associations or habits. The problem of understanding language then became a matter of characterizing the built-in structure of an information-processing system.

Central to the discussion of Miller and his co-authors was the concept of *plan*, a hierarchical process in an organism that controls a sequence of operations. Plans were viewed as similar to the *programs* in Newell and Simon's computational models. Miller and his co-authors argued that the development of cybernetic and computational ideas filled a theoretical vacuum between cognition and action found in previous cognitive approaches such as Tolman's and the Gestalt psychologists. Computational ideas suggested how cognitive structures could produce behavior. In place of the conditioned reflexes of the behaviorists, Miller et al. posited "test-operate-test-exit" units that placed stimuli and responses in feedback units. Plans were postulated to play a role in language consistent with Chomsky's theories, and to play a role in problem solving consistent with the theories of Newell and Simon. Newell and Simon's rule-based theorem prover was quickly generalized into a theory of problem solving applied to several domains (Newell and Simon 1972).

Behaviorist psychology was heavily oriented toward experiments, and the cognitive approach did not abandon this orientation, even though computer simulation now offered another method. Psychologists resumed experimental investigation of proscribed topics such as concept learning, memory structures, and mental imagery. Ulric Neisser's 1967 textbook *Cognitive Psychology* was organized to follow stimulus information inward from the organs of sense, through many transformations and reconstructions, through to eventual use in memory and thought. Three years later the journal *Cognitive Psychology* was started. From Newell and Simon's work, along with the contemporaneous research of Marvin Minsky and John McCarthy, grew the field of artificial intelligence. While some AI researchers prefer an engineering approach that attempts to make computers intelligent independent of how human thinking works, cognitive modeling remains a vibrant approach in AI. In the mid-1970s, the term "cognitive science" was coined to refer to the interdisciplinary effort—embracing psychology, artificial intelligence, linguistics, neuroscience, anthropology, and philosophy—to understand the structures and processes of the mind.

The most successful AI research in the 1960s was a continuation of Newell and Simon's rule-based approach and its extension to the development of the first expert systems. In the 1970s, an alternative approach to processing information arose based on more concept-like structures, the frames of Minsky (1975) and the scripts of Schank and Abelson (1977). The related psychological notion of a *schema* came to play a large role in psychological theorizing. By far the most striking development in cognitive science in the 1980s has been the growth of connectionist models. Connectionism is viewed by some supporters as a revolutionary supersession of cognitivism, and by some critics as a throwback to behaviorism, but I see it as a framework within the cognitivist approach.

Much fuller histories of the development of cognitivism than this brief sketch are available (Gardner 1985; Baars 1986). But let us now turn to analysis of the rise of cognitivism in terms of conceptual change and explanatory coherence.

9.3.2 Cognitivism and Conceptual Change

It might seem that the revival of cognitivism marked a return to commonsense psychology. Cognitivism is certainly closer to everyday psychology than behaviorism, since it allows use of terms like "thinking" in usages not far from their ordinary uses. But cognitivism involves much conceptual innovation that takes it well beyond commonsense psychology or introspectionism. The concept of information as something that can be processed by a person or a computer is a strong extension of the ordinary concept. The concept of *information processing* as applied to the human mind is a major innovation.

Although it has come to dominate cognitive psychology, the information-processing view has not been accepted in all quarters. Ulric Neisser (1976), following Gibson's "ecological" approach to perception, has been critical of computational models. He prefers to emphasize the presence of information in the *world*, advocating an *ecological* approach to the study of cognition. Similarly, the philosophers Jon Barwise and John Perry speak of information as a property of facts or situations (Barwise and Perry 1983). Cognitive science has thus operated with three notions of information: the mathematical, the processing, and the ecological. The ordinary concept of information has thus been differentiated. It makes no sense to ask in the abstract which new concept is correct. Ideally, we would like to have a unified theory that could show how information can be found in (1) representations in minds and computers, (2) the world, and (3) channels of communication between minds and/or computers, and from the world to minds and/or computers. Such a unified theory would go a long way to providing answers to epistemological puzzles about how mental representations can possess content and be meaningful.

Perhaps commonsense psychology has a concept of mental representation, since ordinary people think of beliefs and desires as things "in the head" that relate to the outside world. But the scientific concept of representations in computers as well as humans goes well beyond the ordinary concept. Comparison between humans and computers has allowed theorizing about a variety of complex representations. In artificial intelligence, the two most important kinds of knowledge structures have been rules and frames or schemas. Psychologists have made heavy use of these structures in their cognitive theorizing, as well as a host of other computational ideas: buffers, semantic networks, heuristics, depth of processing, and so on. Thus there are many concepts in the cognitive approach that are new with respect to commonsense psychology and introspectionism, as well as with respect to behaviorism.

The method of cognitivism is different from that of commonsense psychology and introspectionism. Like behaviorists, cognitive psychologists often do carefully controlled experiments. Sometimes in less controlled experiments subjects are asked to think aloud while they solve a problem, but their verbal reports are not taken as veridical mental observations, as the introspectionists would have had it, but as data to be explained by a theory of mental processing. Commonsense psychology is based on unsystematic generalizations about other people and ourselves, and is not based on controlled experiments or explicit models.

Compared to behaviorism, cognitivism adds many concepts for describing mental structures and processes; but it does not delete most behaviorist concepts, although some can be reinterpreted or narrowed in scope. For the high-level cognitive activities of humans, cognitivism employs a different conceptual system than the behaviorists, as summarized in Miller (1965, 20):

> If we accept a realistic statement of the problem [of language], I believe we will also be forced to accept a more cognitive approach to it: to talk about hypothesis testing instead of discrimination learning, about the evaluation of hypotheses instead of the reinforcement of responses, about rules instead of habits, about productivity instead of generalization, about innate and universal human capacities instead of special methods of teaching vocal responses, about symbols instead of conditioned stimuli, about sentences instead of words or vocal noises, about linguistic structure instead of chains of responses—in short, about language instead of learning theory.

But for explaining simpler kinds of behavior in humans and other animals, the behaviorist vocabulary still finds some applications. While denying the universal applicability of concepts such as stimulus, response, conditioning, and reinforcement, cognitivists nevertheless do not deny their usefulness for some kinds of behavior. No one denies that Pavlovian or operant conditioning exists, although it has been reinterpreted in more cognitive terms. According to Rescorla (1988, 153), conditioning "involves the learning of relations among events. It provides the animal with a much richer representation of the environment than a reflex tradition would ever have suggested." Holyoak, Koh, and Nisbett (1989) present a model of conditioning as rule learning rather than mere adjustment of habit strength. Thus conditioning is absorbed into the conceptual system of cognitivism, not deleted. Psychologists no longer postulate Hull's intervening variables for internal stimuli and responses, but these can be understood as a limiting case of the rich sorts of internal representations that abound in cognitivist theories.

How are the new concepts of cognitivism organized? The most significant change in kind-hierarchies brought about by cognitivism is the classification of both mind and computer as information processers. This reorganization has metaphysical as well as scientific importance. Before the development of computers, behaviorists hounded cognitivists with the charge that use of mentalistic concepts was dualistic, implying a separation between mind and body, between the mental and the physical. This charge was fair against introspectionists like Angell, who indeed saw psychology as having a different subject matter from the natural sciences. But no one denies that computers are physical entities: if attribution of internal representations and processes to them is methodologically admissible, then so should be attribution of representations and processes to the brain. Cognitivism brings with it a coalescence of the concepts of mind and computer, using the new, higher concept of information processor.

In commonsense psychology, thinking is a kind of mental activity. For Watson, thinking was reclassified as a response, subvocal talking. In cognitivism, thinking can still be thought of as a kind of mental activity, but for theoretical purposes it is also thought of as a kind of computation. On a weak interpretation of cognitivism, thinking is seen as *analogous* to computation, but the bolder interpretation claims that thinking *is* computation. That is not to say that thinking is computation as performed by a Turing machine or any of the kinds of computers we have available today. We are only beginning to learn what kind of computer the mind/brain might be.

Some concepts of commonsense psychology still do not fit well into the cognitive conceptual system. Cognitivists allow unconscious processes, but there is no established theory of consciousness and its role, if any, in information processing. It is also not clear how to organize concepts describing activities such as *thinking* and *learning*. Rips and Conrad (1989) have offered an analysis of the hierarchical relations of mental activities, claiming that in commonsense psychology the kind-hierarchy and part-hierarchy are the same, running in different directions. For example, planning is a kind of thinking, but thinking is part of planning. Fellbaum and Miller (1990) argue however that conceptual organization based on verbs like "think" and "plan" should not be modeled using the kind-relations and part-relations of nouns. On their analysis, the key hierarchical relations between concepts represented by verbs are *manner-of* and *entails*. Both of these relations form the same bidirectional hierarchy. Planning is a manner of thinking, and thinking is entailed by planning. In response, Rips (1990) defends the view that concepts representing mental activities are organized by kind-relations and part-relations.

To sum up, alterations in kind-hierarchies brought about by cognitivism include coalescence and branch jumping, as mind and thinking become recategorized in computational terms. From a purely conceptual point of view, cognitivism thus has a mildly revolutionary character, since many concepts were added and reorganized, although cognitivism involves little deletion of commonsense and behaviorist concepts. Unlike the seven revolutions in the natural sciences, however, we cannot say that the conceptual change was primarily brought about by virtue of the explanatory coherence of a new unifying theory.

9.3.3 Cognitivism and Explanatory Coherence

I have described cognitivism as an approach, more characterized by methods of experiment, computer simulation, and postulation of mental structures and processes than by a specific theory. Of course, cognitivists would love to have a general theory of mind if they could get one. But there have been few attempts to produce an account of a wide range of cognitive phenomena, and most of these have been explicitly offered as frameworks rather than theories (Anderson 1983; Holland et al. 1986). Newell (1990) argues, however, that psychology is ready for a unified theory, and sets forth a candidate based on the system SOAR.

So what is the relation between the various frameworks of the cognitivist approach and the frameworks of the behaviorist approach? Behaviorist concepts like conditioning have largely been absorbed into cognitivism. But just as relativity theory showed Newtonian mechanics to be of unexpectedly limited application, so cognitivism greatly restricts the explanatory range of behaviorist ideas. Hence the relation between cognitivism and behaviorism is best described as *sublation*, since cognitivism incorporates much of behaviorism while rejecting many of its explanatory claims. Cognitivism also aims to sublate commonsense psychology, because it incorporates, at least potentially, many of the simple explanatory principles of everyday life, while adding a host of computational concepts to the explanatory techniques of everyday life. Cognitive scientists often incorporate these concepts into their everyday explanations, saying things like "he's in a loop" or "she's not sushi-schematic." But cognitivism does not simply add to commonsense psychology, since it occasionally rejects as inadequate the explanations that people give. For example, Nisbett and Wilson (1977) showed cases where people misidentified their own inferential processes, and social psychologists have found that people tend to overuse personality traits in explaining the behavior of others (Nisbett and Ross 1980). Hence we should describe cognitivism as sublating commonsense psychology rather than as incorporating it.

If cognitivism does not contribute a theory whose explanatory coherence led to the demise of behaviorism, why did it come to dominate? The answer, similar to the answer to the question about the rise of behaviorism, is that it appeared to have the promise of greater explanatory coherence. But where behaviorism's greater expected coherence over introspectionism came from the simplicity of its principles, cognitivism's greater expected coherence over behaviorism derived from the much greater potential range of facts explained. Behaviorism's early promise to explain a full range of human behavior never panned out, so it is not surprising that psychologists turned to a different approach with more promising conceptual resources, especially computational ones. Although adoption of a new approach is not characterizable by any process so determinate as found in ECHO, it can nevertheless be rational. Explanatory coherence is based on relations of the sort: *Hypothesis H1 explains evidence E which H2 does not explain.* But evaluation of an approach is based on the more slippery relation: *Approach A1 looks like it will be able to produce a theory to explain E, but approach A2 does not seem powerful enough to produce such a theory.* Such judgments are clearly more subjective and problematic than ones that rely on the exact comparison of existing explanatory theories. But judgments about fruitfulness of different approaches must be made, and they can lead to substantial amounts of conceptual change. Evaluation of approaches is made on the basis of actual and *expected* explanatory coherence. Given the complexity of cognitive phenomena, ranging from vision to language to inference, we should not perhaps expect to have an overarching cognitive theory, since different processes may require different principles. Only further psychological research will determine whether the expectations of cognitivism will be fulfilled.

What strands of explanation are exhibited in psychological explanations? Because computational mechanisms are so important in cognitivism, we can identify the major strands of explanation as causal and analogical. Thinking is explained in terms of structures and processes akin to those found in computers. All the other strands of explanation—deductive, statistical, and schematic—also turn up in explanations in cognitive psychology in minor ways. Behaviorist explanations were different. Since behaviorists like Watson and Skinner tended to avoid postulation of causal mechanisms, their explanations lacked the causal strand, and of course they had no recourse to computational analogies. Since behaviorists often repeatedly applied the same simple patterns of stimulus and response to different phenomena, the major strand of behaviorist explanations seems to be schematic, although there were undoubtedly deductive and statistical components as well.

9.3.4 Connectionism

The most important development in the cognitive sciences in the 1980s was the revival of neural-style computational modeling. In the 1950s, computer simulation of neural networks was a popular line of investigation, but it largely died out with the rise of artificial intelligence in the 1960s.[2] But AI was not so dramatically successful as had been hoped. Limitations were evident in all the approaches advocated by AI's founders, including the rules of Newell and Simon, the frames of Minsky, and the logic of McCarthy. Psychologists and computer scientists began to develop a different style of computational model, consisting of units roughly analogous to neurons connected

[2] Previously in psychology, Thorndike (1949) referred to himself as a "connectionist," but this had nothing to do with neural modeling: his connections were between stimuli and responses.

by excitatory and inhibitory links. Applications to interesting cognitive phenomena, including vision, semantic networks, and language learning were quickly developed (Hinton and Anderson 1981; Feldman and Ballard 1982; Rumelhart and McClelland 1986).

Some philosophers and cognitive scientists have made strong claims about connectionism. We can stereotype two extreme positions:

1. Connectionist learning models, with their flexibility and resemblance to the brain, have made traditional AI models obsolete.
2. Connectionism is merely a bunch of implementation details, relevant to running programs on parallel architecture but suggesting nothing interesting about the functional architecture of cognition.

Both these imperialistic positions block the road to promising inquiries. Advocates of connectionism claim to introduce a new "paradigm" in the Kuhnian sense. There are many Kuhnian senses of that term, and connectionism qualifies in only some of them. Connectionism is clearly not a theory in the sense of an identifiable set of explanatory hypotheses. Rather, it is best characterized as a framework with novel concepts that function within the cognitivist approach. Connectionism is cognitivist in spirit, since it postulates mental representations and computational processes, though different ones from previous frameworks.

I see three central concepts in current connectionist research. The first is connectionist learning in which feedback to the output layers of a network is used to adjust the weights on the links between units, particularly between the hidden units and input or output units (see section 2.4). Although learning of weights through feedback is a useful idea, it is hardly new. According to Rescorla (1988), the delta rule used in back-propagation algorithms is virtually identical to a rule used in accounting for Pavlovian conditioning. Any supposition that such an algorithm is the basis for much human learning would take us back to J. B. Watson. Undoubtedly, more sophisticated learning algorithms will be developed within the connectionist framework, but there are few connectionist systems that model more flexible sorts of unsupervised learning. And the delta rule as found in connectionist models seems to operate more slowly than even classical conditioning, often taking hundreds of trials to learn simple relations.

Why, then, has there been so much excitement about connectionist learning schemes? The answer, I think, lies in the second central concept of connectionism: distributed representations. Although back-propagation methods themselves do not contribute much to our understanding of learning, they produce a kind of representation that is genuinely novel. Instead of a concept such as "horse" being represented by a particular symbol or data structure, the concept can be distributed in the weights between the various units in a complex network. Such distribution provides much flexibility and appears to have some advantages over production rules and frames. Learning and representation are much more intimately related in connectionist models, since distributed representations are produced by learning (Hanson and Burr 1990).

Doubts remain, however, about the representational adequacy of such networks, since they have been used primarily for simple concepts of the sort represented by one-place predicates like "horse." A concept like *rides*, where there is a crucial difference between a man riding a horse and a horse riding a man, requires more expressive power than simple connection between nodes. It is no accident that the implementation of rule-based reasoning in a distributed network can handle only rules with one-place predicates (Touretzky and Hinton 1988). To represent the rules that constitute much scientific knowledge, we clearly need relations and very complex propositions, for example in Newton's laws. Hence I have not used distributed representations in my own investigations of scientific reasoning. The concept of distributed representation is powerful and appealing, but the connectionist framework has a long way to go to catch up to the expressive power of the rule and frame approaches, even if it may win out in flexibility in the long run. As an intermediate strategy, I have found it useful in ECHO to use local representations with units representing whole propositions.

The third major concept of connectionism, the one that has proved most useful in my own research, is parallel constraint satisfaction. Chapter 4 described the advantages of conceiving of theory evaluation as the satisfaction of multiple constraints deriving from explanatory relations, and Keith Holyoak and I have found a similar approach powerful for research on analogy (Holyoak and Thagard 1989; Thagard et al. 1990). Far from just providing different implementation details, the connectionist approach leads one to think of problems in different ways, suggesting different algorithms for solving different difficult cognitive tasks. Connectionism is sometimes reviled as a reversion to behaviorism or associationism, a judgment encouraged by the simple learning algorithms that connectionists tend to use. Cognitivists have to ask: Whatever happened to high-level processes like hypothesis formation? But from the perspective of parallel constraint satisfaction, connectionist models are much more like Gestalt psychology than like behaviorism. Using a tempered holism, ECHO activates units representing complexes of hypotheses, and the same kind of holistic assessment is found in parallel constraint models of analogy, vision, and language. Such models are cognitivist through and through.

If connectionism and traditional AI offered competing theories, we would have to choose between them on the basis of their explanatory coherence. Since, however, they are only different frameworks, we are free to construct a synthesis of valuable concepts from them. Hence I have not hesitated to borrow constraint-satisfaction ideas from connectionism, while retaining traditional AI representations of concepts and rules. Connectionism is clearly an important research program, offering new ways of thinking about learning as well as new ways of implementing programs in parallel hardware. But dramatic advances in connectionist algorithms will be needed to duplicate the success of traditional AI programs in such tasks as unsupervised discovery. Connectionism should therefore be understood as an important development in cognitive science, adding to its representational and computational resources. But there is no reason to expect a connectionist revolution in psychology, even though the influence of connectionist ideas in psychological theorizing will probably continue to grow.

A much more radical view of connectionism is offered by Paul Churchland (1989). He sees it as the computational part of a neuroscience research program that will eventually eliminate folk psychology and much of current cognitive psychology. Just as physics has abandoned theories of phlogiston and caloric, so will computational neuroscience lead to the replacement of folk-psychological and current cognitive theories. From Churchland's perspective, the use by ECHO of nodes representing propositions is an archaic variant of traditional AI and neglects the possibilities of the more radical connectionist ideas based on distributed representations.

I have two responses, one based on the inadequacies of current work on neural and distributed representations, and the other based on the relations between different kinds of representations. First, as I argued a few paragraphs back, current work on distributed representations is much too primitive to model thinking involving scientific theories. No one has the vaguest idea how to train up a network that would possess anything like Darwin's theory of evolution or Newtonian mechanics, and a neural understanding of such theories is even more remote. Churchland (1989, 188) proposes that a theory is a global configuration of synaptic weights in a neural network, and accepts the implication that even the simplest of animals possess theories. Explanation is the elicitation of a pattern of activation of hidden units representing a prototype, and inference to the best explanation is activation of the most appropriate prototype. These proposals have an appealing simplicity, but their explanatory accomplishments will be meager until the computational power of systems with distributed representations expands to encompass real scientific theories. At the very least, there is a major difference in degree between the theories and explanations of scientists and the cognitive structures of cockroaches. We saw in section 5.3.1 that prototype application is only one of several aspects of explanation. Although I agree with Churchland that not all thought is propositional and linguistic, I would argue that much high-level human cognition has a propositional component not found in simpler animals.

My second response to Churchland's rejection of local representations such as those used by ECHO is that they can be viewed as an approximation to distributed ones, just as the distributed representations found in current computational models are only an approximation to those found in actual neural systems. Even if the deepest explanations of scientific thinking eventually turn out to come from a neurocomputational level of distributed representations, models employing local representations may still be very useful. McCauley (1986) has pointed out a crucial flaw in Churchland's analogy between the posits of folk psychology and eliminated concepts such as phlogiston. Deletion of such concepts is the result of competition between theories at the same level, as we saw with the oxygen and phlogiston theories. But the history of science reveals no precedent for theory replacement or elimination in cases such as cognitive psychology and neuroscience which operate at different levels. My methodological recommendation is that research should proceed at all relevant levels, investigating the power of local and distributed representations and their relation to neural structures. We can expect connectionist ideas to aid in the continued sublation of folk psychology by cognitivism, not to bring about the complete replacement of everyday theories of mind.

9.4 OTHER DEVELOPMENTS

A full canvas of movements in psychology that have been called revolutionary would have to deal with two figures who are undoubtedly giants of twentieth-century investigation of the mind: Freud and Chomsky. The "Freudian revolution" introduced numerous concepts into psychology and ordinary culture: unconscious desires, infantile sexuality, the Oedipus complex, repression, the Id and Superego, and so on. Freud proposed a division of the mind into parts not found in commonsense psychology or behaviorist or cognitivist theories, and I expect that a thorough search would find interesting changes in kind-relations as well as part of the Freudian framework. On an anecdotal level, the Freudian systems possesses much explanatory coherence, but there is little experimental evidence to support the vast edifice. Freudian psychology had a large influence on clinical psychology, but even that is waning rapidly. So despite the considerable influence of Freudian ideas, even on behaviorists like Watson, it seems misleading to speak of a Freudian revolution in psychology.

The Chomskyan revolution in linguistics took place in the 1950s and 1960s with the replacement of behaviorist theories of language by Chomsky's view that a system of universal grammar is an innate mental structure in humans. It should be possible to analyze Chomsky's polemics against previous theories held by psychologists and linguists in explanatory coherence terms. But I shall attempt neither that task, nor analysis of the conceptual changes required by Chomsky's theory, nor assessment of the place of his current theories in cognitive science.

Other areas of the social sciences—politics, economics, sociology, anthropology—have undoubtedly undergone conceptual changes in their developments, although like psychology they have been short of unifying theories with much explanatory coherence. It would be interesting to interpret figures such as Marx, Compte, and Keynes in the terms used here. I am happy, however, to leave to other investigators the question of whether the account of conceptual change offered in this book applies to other developments in the social sciences.

9.5 SUMMARY

While psychology has seen much conceptual change in this century, with the replacement of the introspectionist and commonsense conceptual system by behaviorism, and the sublation of behaviorism by cognitivism, it has not had revolutions of the sort so important in the natural sciences. Behaviorism and cognitivism involved abundant conceptual change, including concept deletions and conceptual reorganization involving kind-relations. But they are best characterized as approaches rather than theories, and their ascent depended more on estimates of future explanatory coherence than on evaluation of the explanatory coherence of specific theories.

CHAPTER 10

Conceptual Change in Scientists and Children

UNDERSTANDING the growth of scientific knowledge requires a theory of conceptual change. But the topic of conceptual change has importance that goes beyond the philosophy and psychology of science. Developmental psychologists have been keenly interested in questions of conceptual change as part of their attempts to understand the growth of knowledge in children. Some psychologists have argued that children's learning is like scientific development in that it cannot be described merely in terms of accretion of new beliefs added to old: it involves rejection of previously held beliefs and substantial conceptual change. In addition, researchers interested in science education of students from grade school through college have considered whether students acquiring successively more complicated scientific theories need to go through the same kind of conceptual changes undergone by the scientists who developed those theories.

Most comparisons of scientists and children by developmental psychologists and science educators have been limited by the inadequacy of available theories of conceptual change in science. If the much more detailed account of conceptual change in earlier chapters is acceptable, then richer comparisons of scientists and children become possible. We can contrast conceptual change in scientists during revolutionary periods in the history of science with conceptual change in children and in students learning science. There are points in common between conceptual change in the history of science and conceptual change in ordinary learners, but there may also be important differences. Ordinary learning is not simple accretion, but it may not be as systematic or as revolutionary as conceptual revolutions in science.

Before embarking on a comparative discussion of conceptual change in children, I shall review the results derived so far on scientific revolutions. Paying close attention to the history of science has many advantages, especially compared to using made-up cases, but there is always the risk of getting lost in the details. We have examined seven revolutions in the natural sciences, and two quasi-revolutions in psychology, and it is time to draw the discussion together. I have not tried to define the concept of a conceptual revolution, but have showed similar patterns of conceptual change arising in various conceptual revolutions. The next section presents a series of comparative tables that summarize the findings about conceptual change and explanatory coherence discussed in previous chapters. The summaries are no substitute for the theoretical and historical discussions in Chapters 3–9, but provide a convenient way to present the character of the different conceptual revolutions.

10.1 COMPARATIVE SUMMARY OF SCIENTIFIC REVOLUTIONS

10.1.1 Discovery and Conceptual Change

A cognitive/computational approach to understanding of science has the great advantage over traditional philosophical or sociological approaches that it can start to specify in detail the psychological mechanisms that lead to important scientific discoveries. In different revolutions, however, we see different kinds of discovery processes operating. Chapter 3 described a range of discovery methods as data-driven (generalized from observations and experiments), explanation-driven (abductive), and coherence-driven (formed to overcome contradictions). Table 10.1 presents examples of the concepts and hypotheses that fall under these three categories of discovery. The revolutions are listed in the order in which they were discussed in previous chapters. A blank entry does not indicate that there were no discoveries in that category for a particular revolution, only that the category is comparatively unimportant.

What we learn from Table 10.1 is that a variety of different discovery methods were instrumental for the different revolutions. Data-driven discovery of empirical generalizations played a large role in the work of Lavoisier, Darwin, the geologists whose sea explorations led to Hess's hypothesis of seafloor spreading, and behaviorists. In contrast, for Copernicus and Einstein, the major impetus to discovery came from incoherencies in existing views.

Table 10.1
Modes of Discovery

Revolution	*Data-driven*	*Explanation-driven*	*Coherence-driven*
Lavoisier	weight gain	oxygen	
Darwin	fossil patterns	natural selection	
Geology	sea ridges	drift, plates	
Copernicus		Earth's rotation	Sun at center
Newton		force of gravity	
Einstein			speed of light
Quantum	Planck's law	light quantum	uncertainty principle
Behaviorist	law of effect	intervening variables	
Cognitivist	magical number 7	mental representations	

Table 10.2
Kinds of Conceptual Change

Revolution	*Concepts added*	*Concepts deleted*	*Branch jumping*	*Tree switching*
Lavoisier	oxygen	phlogiston	metal APO calx gold AKO element	
Darwin	natural selection	divine creation	humans AKO animal	kind: historical
Geology	plate	shrinking earth		continent & seafl
Copernicus			earth AKO planet	
Newton	force of gravity	vortex	motion AKO state	
Einstein	relativistic mass	aether	mass and energy equivalent	space-time
Quantum	light quantum		wave and particle related	indeterminacy
Behaviorist		mind	thinking AKO response	
Cognitivist	information		thinking AKO computation	

Note: AKO means "becomes a kind of." APO means "becomes a part of."

Most revolutions have a large explanation-driven component, in which concepts and hypotheses are generated to explain puzzling facts.

Table 10.2 compares the conceptual developments in the various revolutions. Most revolutions involve the addition and deletion of concepts, as well as branch jumping in which the kind-relations and/or part-relations are reorganized. Recall that branch jumping is a more drastic kind of conceptual change than simply adding a new kind-relation or part-relation, since it requires rejecting a previous relation. Even less commonly, tree switching occurs in which the kind-hierarchy or part-hierarchy is fundamentally changed.

10.1.2 Explanatory Coherence

Explanatory coherence is the main mechanism of theory adoption in conceptual revolutions, but different aspects of explanatory coherence are important in different cases of theory replacement. Table 10.3 compares the revolutions with respect to what contributed most to the greater explanatory coherence of the new theory. Once again, the table entries are cryptic and the reader should refer to the relevant chapter for the details. For most revolutions, the explanatory breadth of the new theory—how much it explains—is the largest contributor to its explanatory coherence. Simplicity usually plays a role, large or small, but analogy is a less frequent contributor to explanatory coherence. The hierarchy column concerns whether the hypotheses in a theory are themselves explained; the best example of this is Darwin's Malthusian derivation of natural selection. Contradictions play a role in explanatory coherence when the opposing theory either is internally contradictory or explains negative evi-

Table 10.3
Contributions of Explanatory Coherence

Revolution	*Breadth*	*Simplicity*	*Analogy*	*Hierarchy*	*Contradictions*	*Prospects*
Lavoisier	large	small			small	
Darwin	large	small	small	small		
Geology	large			small		
Copernicus	small	large			small	
Newton	large	large		small	small	
Einstein	large	small			small	
Quantum	large				small	
Behaviorist	small	small				large
Cognitivist	small		small		small	large

dence. Finally, prospects were important for the quasi-revolutions in psychology, which were more driven by expectations of explanatory coherence than by actual explanatory coherence.

As section 5.3 described, the kind of explanation that goes into judgments of explanatory coherence varies from field to field and theory to theory. Table 10.4 summarizes the strands of explanation that were most prominent in the analyses of the explanatory coherence of the different theories. Explanations in physics have been more strictly deductive than those in other fields. Darwin's explanations often had a statistical flavor, even though he was not aware of it. Darwin and Newton are marked as being more schematic in their explanations because they used recurring patterns of explanation more than the others. Postulation of causal mechanisms appears important in most cases. More careful scrutiny of explanations offered in relativity theory, for example, might lead to additional entries.

Table 10.4
Strands of Explanation

Revolution	*Deductive*	*Statistical*	*Schematic*	*Analogical*	*Causal*
Lavoisier	small		small		large
Darwin	small	small	large	small	small
Geology	small				
Copernicus	large		small		
Newton	large		large		small
Einstein	large			small	
Quantum	large	large			
Behaviorist	small	small	large		
Cognitivist	small	small	small	large	large

Table 10.5
Relations between New and Old Systems

Revolution	*Incorporate*	*Sublate*	*Supplant*	*Disregard*
Lavoisier			yes	
Darwin			yes	
Geology			yes	
Copernicus			yes	
Newton			yes	
Einstein		yes		
Quantum		yes		
Behaviorist			yes	
Cognitivist		yes		

Finally, Table 10.5 summarizes the conceptual and explanatory relations between the old and new theories or approaches. Chapter 5 described four possibilities: incorporation, in which the old theory is merely absorbed into the new; sublation, in which rejection and reorganization of concepts takes place even though the explanatory hypotheses and successes of the old theory are absorbed into the new; supplantation, in which the old theory is summarily rejected; and disregard, in which no attention is paid by proponents of the new theory to the old theory. Most nonrevolutionary developments in science are best characterized as incorporation, and the developments in nonscientific domains and in some social sciences are often best described as disregard: previous views are simply ignored. But revolutions in the natural sciences tend to fall under the categories of supplantation and sublation. Strikingly, physics has undergone sublations as well as supplantations.

10.1.3 Revolutions and Scientific Knowledge

Tables 10.1–10.5 summarize some of the evidence for the six theses on scientific revolutions put forward in Chapter 1:

1. Scientific revolutions involve major transformations in conceptual and propositional systems.

2. Conceptual systems are primarily structured via kind-hierarchies and part-hierarchies.

3. New theoretical concepts generally arise by mechanisms of conceptual combination.

4. Propositional systems are primarily structured via relations of explanatory coherence.

5. New theoretical hypotheses generally arise by abduction.

6. The transition to new conceptual and propositional systems occurs because of the greater explanatory coherence of the new propositions that use the new concepts.

The last thesis makes it appropriate to label my view as an *explanationist* theory of the growth of scientific knowledge. Explanationism can be distinguished from two more familiar approaches to understanding science, empiricism and rationalism. Empiricists emphasize the role of sense experience in guiding the development of science, while rationalists explain science in terms of general conceptual considerations. Empiricism typically degenerates into skepticism about the truth of scientific theories, since sense experience alone does little to determine the generation and acceptance of scientific theories. Rationalism tends to degenerate into irrationalism, since lack of a mind-independent component to reasoning renders arbitrary what theory is adopted by scientists.

Explanationism combines the most plausible aspects of empiricism and rationalism. The principle of data priority that is part of the theory of explanatory coherence of Chapter 4 asserts that the results of observation and experiment should indeed contribute in a special way to theory evaluation. But there is much more to evaluation than simply finding the best "fit" with what is observed. Theories are enmeshed in conceptual systems, just as rationalists have argued, and such systems are not open to simple empirical challenges. Nevertheless, a few straightforward principles of explanatory coherence can account for the theoretical choices made by scientists. Does explanationism also apply to the development of knowledge in children?

10.2 CONCEPTUAL CHANGE IN CHILDREN

10.2.1 Are Children Like Scientists?

Developmental psychology was long dominated by the theories of Jean Piaget, who described children's development in terms of a fixed set of cognitive states. Research has undermined his claims that children universally move from preoperational thinking to concrete operations to formal operations. Instead, a growing number of developmental psychologists argue that changes in children are better described in the same terms that describe the growth of scientific knowledge: conceptual change and theory replacement. By reviewing some of the most important recent studies of children's intellectual development, we can compare the growth of knowledge in children with the growth of knowledge in scientists, taking for granted the account of scientific change given in previous chapters. The central questions are:

1. Do children undergo the same kinds of conceptual change as scientists?

2. Do children have theories that provide unified explanations of observed facts?

3. If so, when children acquire theories, do these theories replace previously held theories?

4. If so, is the replacement of children's theories the result of considerations of explanatory coherence?

Although definitive answers to these questions are not yet possible, a growing body of experimental and theoretical research in developmental psychology is highly suggestive. Carey (1985) describes the development of children's biological concepts as analogous to conceptual change in science. Keil (1989) argues that coherent belief systems are critical to understanding the nature of children's concepts and how they develop. Vosniadou and Brewer (1987, 1990) have collected extensive data on the development of children's beliefs about the shape of the earth. They contend that children construct their own models and that replacement of these models with adult conceptions of a spherical earth is analogous in many respects to theory change in the history of science. Wellman (1990) argues that children as young as age three have a commonsense theory of mind that differs from the theories of older children and adults. Chi (1991) maintains that students' learning of scientific concepts is impeded by their need to shift concepts across ontologically distinct categories, although she rejects the view that acquisition of scientific theories by students involves their rejection of prescientific ones. How do these developmental views fit with the theory of conceptual change presented in earlier chapters?

10.2.2 Changes in Children's Concepts

In Chapter 3, I distinguished numerous kinds of epistemic change. First I distinguished conceptual change from belief revision, in which relations between concepts are established or rejected without deeply affecting the concepts. Belief revision involves either the addition or deletion of beliefs. Conceptual change comes in varying degrees:

additions of concepts;
deletion of concepts;
simple reorganization in the kind-hierarchy or part-hierarchy, in which new kind-relations or part-relations are established;
revisionary reorganization in the hierarchies, in which old kind-relations or part-relations are replaced by different ones; and
hierarchy reinterpretation, in which the nature of the kind-relation or part-relation that constitutes a hierarchy changes.

Carey (1985) suggests that children undergo conceptual change analogous to the radical kinds of conceptual change described by historians of science such as Kuhn. Her "paradigm cases" of conceptual change are the coalescences, differentiations, and alterations in ontological commitments that occur in such scientific transitions as the one from Aristotle to Galileo. Carey contends that the acquisition of biological knowledge by children between the ages of 4 and 10 likely involves theory change akin to that found in science. She says that children shift from an animist theory, in which the sun and trees are counted as alive, to a set of biological theories explaining facts that the animist theory would explain in terms of intentional activity. A very young child might count the sun as alive because it gives light, and say that trees get the idea of blooming from other trees, but by the age of 10 children have acquired enough biological knowledge to reject their old beliefs. Ten-year olds understand death, reproduction, gender, digestion, circulation, and respiration in terms of internal bodily processes. Carey describes a battery of psychological experiments concerning children's concepts of life, animals, and body parts in support of her interpretation of the development of children's knowledge as akin to theory change in science.

Let us compare Carey's description of children with my account of conceptual revolutions in science. Between the ages of 4 and 10, children clearly acquire many new biological concepts, for example referring to body parts that they did not previously know about. Abandonment of concepts occurs too: for preschoolers, fairies and goblins are as real as horses and buses; children cease to use such concepts when they learn that the entities referred to do not exist. In these cases, however, concept deletion does not seem to be part of theory replacement, since there is no obvious theory that children use to replace their tales of fairies and goblins.

Carey's most plausible case for conceptual change in children concerns the concept *living thing*. Children distinguish between *alive* and *dead*, but not between *alive* and *inanimate*. Whereas 10-year olds are able to class animals and plants as alive, in distinction from inanimate objects, younger children's categories are much more blurred. Part of the knowledge reorganization that occurs as children acquire more biology is the coalescence of *animal* and *plant* into the superordinate concept *living thing*. Carey compares this development to Galileo's collapse of Aristotle's distinction between *natural* and *violent* motion, but in Chapter 3 I argued that coalescence and collapse are distinct kinds of conceptual change. More apt comparisons for the acquisition of *living thing* are with Maxwell's coalescence of electricity and magnetism into *electromagnetism* or cognitivism's coalescence of mind and computer into *information processor*. Coalescence is a less extreme sort of conceptual change than collapse and branch jumping; these changes require rejection of accepted relations, not just addition of new ones. Carey presents no evidence that children undergo the kind of conceptual change that I have classed as revisionary. Nor does it appear that children undergo any sort of hierarchy reinterpretation (tree switching): they learn many more kinds of things, but the notion of *kind* does not change.

According to Carey, children's concepts of *animal* and *person* do not undergo changes as fundamental as *living thing*, although children do learn much new information about animals and persons, for example that all animals eat, breathe, and reproduce. The kind of epistemic change involving these concepts appears to be best described as belief addition, the least extreme form of epistemic change. Carey's studies do not suggest that children have to abandon previously held beliefs about animals and people as their knowledge grows.

Children's part-hierarchies also expand substantially as their knowledge grows. Young children think that a painted skunk can become a raccoon, but older ones know that deeper biological considerations prevent such an easy transformation. Children begin with little knowledge of internal organs, but acquire knowledge of body parts and the processes that they involve. From Carey's description, however, it appears that the development here is accretional: decomposition of the body into parts and belief additions concerning the functions of those parts. Children undergo nothing like the branch jumping that happened with part-relations in the chemical revolution or the reinterpretation of the part-hierarchy in terms of space-time that occurred with relativity theory. Compared to scientific revolutions, conceptual change as described by Carey is modest: some nonrevisionary conceptual reorganization without much rejection of previously held beliefs.

Chi (1988) offers a different view from Carey's of how children develop away from animism. She postulates an incremental process in which children come less and less to associate being alive with being able to move and more and more to associate it with biological functions. Additional concepts and links are acquired over time resulting in a different pattern of emphasis, but the old association between being alive and moving is not actually rejected, although it ceases to be accessed from memory. Chi (1991) also thinks that no actual rejection takes place when students acquire new theories in physics, even though the new theories have very different conceptual structures from everyday theories. She argues that learning physics requires shifts of concepts across ontologically distinct categories such as *material substance* and *event*. For naive students, force, light, heat, and current are kinds of substances, but physics students must learn to reconceptualize them as fields, which she describes as belonging to a different ontological category. Learning physics would then require substantial branch jumping in the sense of Chapter 3, with central concepts moving from one part of the kind hierarchy to another.

Like Carey, Keil (1989) has studied the development of children's biological knowledge; he sees it as involving theories, but not conceptual revolutions. Preschoolers verbalize their concepts as largely associated with characteristic features, but older children express their concepts with utterances that cite more central, defining features connected with an understanding of underlying causal mechanisms. For example, school-age children know that a kitchen pipe can be transformed into a flute, but that surgical changes cannot turn a sheep into a goat. On Keil's account, concepts develop by becoming more and more integrated with biological theories, but he does not describe any of the sorts of conceptual revisions that occur in scientific revolutions.

Vosniadou and Brewer (1987, 1990) have investigated the development of children's knowledge of astronomy. They describe how children, who are routinely told that the earth is round, have difficulty reconciling this claim with their other beliefs and observations. Although answers from most fifth-grade students in the studies suggested that they thought the earth was a sphere, most first-graders were found to have different views. Children are told by parents and teachers that the earth is round, but this contradicts their entrenched belief, based on observation, that the earth is flat. Hence children ingeniously develop models that reconcile these two beliefs. The most popular view among first-graders is the *dual earth* model in which children believe that there are two earths, a flat one on which we live and a round one that is up in the sky. Some children think that the earth consists of two hemispheres: a lower one on which people live and an upper one that has the sky covering the lower one like a dome. On this view, the earth is a sphere, but we live inside it, not on top of it.

Vosniadou and Brewer characterize the children's transition to a spherical earth as conceptual change and theory replacement, but what conceptual change takes place? At first glance, the epistemic change that occurs as children learn more about the earth seems best characterized as belief revision: children abandon the belief that the earth is flat and accept the proposition that it is a sphere. Vosniadou and Brewer's children do not, as far as I can see, add new concepts or delete old ones. Nor are changes in kind-relations evident in Vosniadou and Brewer's descriptions of the children's views. There seem, however, to be interesting changes in part-relations required for a child to move from some models to a full spherical model. Children who believe that people live *inside* the spherical earth clearly have beliefs about the parts of the earth different from older children who see the earth as a solid sphere. Thus for some children, at least, belief revision toward a spherical model involves changes in part-relations. Notice, however, that the part-relations concern a particular planet, the earth, not a general concept.

Even before children learn much biology, physics, or astronomy, their beliefs about human psychology undergo development. On the account of Wellman (1990), children have acquired by the age of three a theory of mind based on beliefs and desires (see also Astington, Harris, and Olson 1988). This theory is an explanatory framework that differs from what is found in younger children, who have a simpler psychology based on desires alone. According to Wellman, children older than three develop a more sophisticated theory in which the mind becomes more active and interpretive, in contrast to the three-year old's view of the mind as a passive container of copies of the external world. The next section discusses Wellman's view that the children possess a theory of mind, but for now I note that Wellman does not describe the development of new psychological theories in children as involving conceptual revision of the sort found in scientific revolutions. Children undoubtedly acquire new psychological concepts as they learn more about people, but on Wellman's account it does not seem necessary for them to overturn previous concepts and their kind-relations or part-relations.

What can we conclude about conceptual change in children from these interesting studies? It seems that changes in concepts from astronomy, biology, and psychology are not of the extreme sorts that attend scientific revolutions. On the other hand, learning physics may involve conceptual reorganization (branch jumping) akin to what is found in major conceptual revisions. Hierarchy reinterpretation does not seem to occur in children's development. It would be premature, however, to reach any firm conclusions about kinds of conceptual change that do *not* occur in children, since new research may well discover that children are in fact undergoing changes that existing studies have not detected.

10.2.3 Children's Theories

The questions remain: do children have theories, does conceptual change occur by theory replacement, and is theory replacement the result of considerations of explanatory coherence? An affirmative answer to each question is a precondition of an affirmative answer to the succeeding one. Some researchers would offer negative answers. According to diSessa (1988), children do not have coherent theories, only loose, unsystematic, but interconnected concepts that do not fit into a set of principles of the sort needed to drive a judgment of explanatory coherence. Deanna Kuhn and her colleagues (1988) question whether children distinguish between evidence and hypotheses, a precondition of judgments of explanatory coherence.

Wellman (1990) makes a thorough case for the claim that even three-year old children possess a theory of mind. They possess a system of beliefs and concepts concerning beliefs and desires that enables them to provide causal explanations of a range of behaviors. Like scientific theories, these systems involve the postulation of nonobserved entities that figure strongly in the causal explanation of what is observed. Brewer and Samarapungavan (1991) have systematically criticized arguments that children do *not* have theories of the natural world, and shown in particular how children's knowledge of astronomy includes components very much like scientific theories, providing explanations for phenomena such as the day/night cycle. It seems, therefore, that at least some of children's cognitive apparatus can be construed as theories. My account of concepts in Chapter 3, which described them as enmeshed in rules, is compatible with the increasing emphasis by developmental and cognitive psychologists on the interrelations of concepts and theories: rules can be used to express causal relations that provide explanations. It is one thing, however, to grant that children have biological and psychological theories somewhat similar to scientific ones, and another to suppose that theory development in children is similar to the kinds of theory replacements that occur in science.

Carey (1985) assumes that the development of biological knowledge in children is a case of theory replacement, in which children reject animism in favor of their new biological theory. It is not surprising that Carey's studies do not detect such transitions, since they compare young children and older children, understandably avoiding the laborious task of tracking children throughout their development. But Keil (1989) questions Carey's view that the development of biological knowledge in children involves the abandonment of a previously held behavioral/psychological theory. Rather than a clash between competing theories, Keil sees children as undergoing elaboration of a rudimentary biological theory that they already possess. Learning is then theory acquisition, not theory replacement. Wellman's (1990) view of the development of children's theory of mind seems similarly accumulative.

On the other hand, Vosniadou and Brewer's subjects must require substantial belief revision to move from some of the stranger models of the earth to the adult spherical model. The difficulty involved in this revision is not captured by considering only conceptual organization into kind-relations and part-relations (W. Brewer, personal communication). Giving up the belief in a flat, motionless earth for belief in a spherical, moving earth is a momentous change that involves no kind-relations. What makes it momentous is apparently that these beliefs are tied in with so many other beliefs. Using the term introduced in Chapter 4, we can say that they possess a great degree of explanatory entrenchment. But is their revision best described as theory replacement of the sort that can occur as the result of explanatory coherence? As with Carey's subjects, it is impossible to tell from the data as presented. Without tracking the first-graders through their subsequent epistemic changes, we cannot determine whether children's old models come into conflict with new models of the solar system and lose out because of the greater explanatory coherence of the latter. Perhaps instead the old models simply fade into the background.

Here are two schemas for what may be happening in children:

1. Children are like scientists. They start off, for example, with an animist theory of why things behave as they do, but when they are taught more about biological and physical mechanisms they consciously or unconsciously appreciate that these accounts have greater explanatory coherence than their old theories and therefore abandon animism.

2. Children simply acquire biology and physics. They do not see these as competitors to their previous animist beliefs, which were not held explicitly or strongly in the first place. Animism is not a set of beliefs that are rejected because they have less explanatory coherence than the new ideas being taught; no comparison is every done. Rather, what occurs is that the new biological knowledge is acquired and the old beliefs are simply forgotten, or reserved for use in everyday nonacademic contexts.

Theory replacement in schema 1 is like having a set of new clothes that you decide are much better than your old ones, so you give the old ones to the Salvation Army. On schema 2, however, without making any general comparison and evaluation, your old clothes simply get pushed into the back of your closet where they may or may not continue to be used. Although Carey prefers to apply to children the first schema that fits revolutionary science, her data appear equally consistent with the back-of-the-closet model. Chi (1991) contends that when students learn theories in physics, they never abandon their old naive theories, but continue to use them in everyday life. More investigation is required to determine whether children undergo theory replacement.

One might wonder whether children have the intellectual capacity to prefer theories on the basis of their explanatory coherence. Samarapungavan (1990) found that elementary school children tended to prefer theories that explain more, lack ad hoc explanations, and are consistent internally and with the evidence. Preferences based on the first two dimensions increased as children got older. This study does not show that children used theory replacement based on explanatory considerations as the vehicle for conceptual changes; but it is encouraging that they have some sensitivity to some of the factors that explanatory coherence theory says are crucial for theory evaluation.

Like the question of how children's concepts change, the questions concerning theory replacement cannot be conclusively answered. Using what is currently known, Table 10.6 summarizes the comparisons of conceptual change in children with conceptual change in scientists. Further research and reconstruction of children's conceptual and explanatory systems may provide more definite answers to the questions posed by the table. These answers may not be universal: it is possible that children's knowledge of different domains develops in different ways. Psychology and biology may be more cumulative than physics, in which students encounter concepts like *field* alien to everyday life. It seems, however, that children's new theories will much more fre-

Table 10.6
Comparison: Science and Children

Kind of Change	*Scientific Revolutions*	*Children*
Concepts added	oxygen, etc.	living thing, etc.
Concepts deleted	phlogiston, etc.	fairy, goblin
New kind-relations	gold AKO element	animal AKO living thing
New part-relations	metal APO calx	spleen APO body
Branch jumping	gold AKO element	force AKO field
Tree switching	historical kinds	?
Explanatory coherence	yes	sometimes
Systematic theories	yes	somewhat
Relation of theories	supplant, sublate	incorporate, disregard?

*AKO means "becomes a kind of." APO means "becomes a part of."

quently incorporate or disregard their old ones, rather than supplant or sublate them. A valuable role for teachers can be to ensure that children become aware of how theories that they have or should have outgrown conflict with what they are being taught in school. Some researchers in science education are currently experimenting with ECHO as a teaching tool that may help in this regard.

10.2.4 The Growth of Knowledge in Children and Scientists

In biology, the thesis that "ontogeny recapitulates phylogeny" means that the development of the individual organism corresponds to the evolutionary development of the whole species. For example, the human embryo has a vestigial tail, a remnant of our evolutionary past. In developmental psychology, "ontogeny recapitulates phylogeny" means that the development of knowledge in individual children corresponds to the development of knowledge in the history of science. Support for the developmental version of the thesis has been drawn from analogies between, for example, the child's transition from holding a naive, Aristotelian view of motion to, eventually, with enough education, a Newtonian view. We must distinguish several recapitulation theses:

1. Content recapitulation: the child goes through historical stages similar to those undergone by scientists;
2. Structure recapitulation: the child undergoes similar kinds of conceptual change and belief revision;
3. Mechanism recapitulation: the child's mechanisms of epistemic change are similar to those of the scientists.

From the perspective of the history of science, the first thesis is implausible. Children's physical theories are not worked out enough to be called Aristotelian or Ptolemaic. Most obviously, children do not, as far as anyone has noticed, have a phlogiston theory of combustion that has to be replaced by an oxygen theory. Although there may be occasional cases where there is some similarity between children's views and historical theories, for example in impetus theories of motion (McCloskey 1983), in general there is not much similarity between the inchoate notions of children and even early scientific views.

Structure recapitulation is somewhat more plausible, since we saw that Carey's subjects underwent some simple kinds of conceptual reorganization involving coalescence, and Vosniadou and Brewer's subjects had some interesting changes in part-relations. Chi's review found evidence for branch jumping in physics education. There is no developmental evidence yet that children undergo the most revolutionary change, hierarchy reinterpretation, but perhaps such evidence will appear in future studies.

Mechanism recapitulation may be partly true. I would not be surprised if the mechanisms by which children build up conceptual systems are similar to the mechanisms by which scientists do so. There is little evidence on the question, but I see no reason to doubt that the concept-generation mechanisms by which scientists acquire new concepts and new kind-relations and new part-relations are substantially different from what occurs in children. More problematic is the claim that children's theoretical development is governed by the kinds of explanatory-coherence evaluations that occur in science.

Even if children do not use ECHO-like mechanisms in their belief revisions, however, ECHO may still have relevance for developmental and educational psychology. Older students may well be capable of the judgments about explanation and evidence that are needed to drive a transition between belief systems similar to what ECHO produces. The best evidence so far available is Michael Ranney's ECHO simulation of his experimental studies of college students revising their beliefs in physics (Ranney and Thagard 1988; see also Ranney 1991). Moreover, there is the possibility that even if explanatory coherence analysis is not fully developed in children, it may be teachable (Bereiter and Scardamalia 1989).

10.3 PROJECTS FOR UNDERSTANDING SCIENCE

The main conclusion of my discussion of conceptual change in children is that much research remains to be done to pin down the kinds of conceptual change that occur as children learn. Of course, the study of conceptual change in scientists is also far from complete. To point to some of the needed research, I now list numerous projects that should illuminate conceptual change within the general perspective adopted in this book. The first set of projects concerns children and science education, while the second set involves cognitive analyses of the history of science.

10.3.1 Children and Education

Conceptual change in students and children may not be as revolutionary as the most extreme kinds of epistemic change in science. But it is possible that extreme changes have not been noticed because no one has looked in the right places. Let me suggest an agenda for developmental and educational research that would exploit the ideas about conceptual change that have arisen from my discussion of scientific revolutions.

1. Reexamine psychological studies of children's development to identify their conceptual systems and changes. Which of the kinds of epistemic change described in Chapter 3 occur in children? In particular, is there any branch jumping in conceptual hierarchies outside physics and is there any hierarchy reinterpretation?
2. Investigate whether explanatory coherence plays a role in children's belief revision and rejection of old conceptual systems. Does theory competition actually occur, or do children merely incorporate old beliefs or ignore them?
3. Determine whether children and students can be taught a greater sensitivity to explanatory coherence issues, and whether this sensitivity can lead them to learn new scientific theories more readily.

Answers to questions such as these will not only increase our understanding of how children's knowledge develops; they may also improve our success in educating children and older students to a more advanced understanding of science.

10.3.2 Scientific Knowledge

My approach to understanding the development of scientific knowledge involves the following steps. First, identify an interesting theory. Second, analyze the conceptual system that contains the theory, paying special attention to kind-relations and part-relations. Third, consider what computational mechanism might have produced the components of the conceptual system. Fourth, analyze the explanatory coherence of the theory compared to its competitors. Although I have provided many illustrations of this method, the analyses have not been complete, even for the cases that I have considered in most detail. Here are some historical and theoretical projects concerning how scientists think, continuing the list involving children.

4. Analyze the conceptual systems of the opponents to Lavoisier's oxygen theory, such as Kirwan and Priestley. Trace the development of the phlogiston theory while its proponents attempted to save it from Lavoisier. Determine whether Kirwan's opposition to the oxygen theory and his eventual acceptance of it can be understood in terms of explanatory coherence.

5. Simulate the transition from phlogiston to oxygen (cf. O'Rorke et al. 1990): program a problem solving system that starts with the phlogiston theory, develops or is given an oxygen theory, and adopts the latter because of its greater explanatory coherence.

6. Analyze the conceptual systems and explanatory coherence of Darwin's critics (Hull 1973). Do the same for modern-day creationists.

7. Analyze conceptual change, theory development, and explanatory coherence in Harvey's theory of the blood circulation. Pay special attention to Harvey's analogy between the heart and pumps.

8. Analyze the processes underlying the discovery by Hess and Dietz of seafloor spreading.

9. Analyze the conceptual and explanatory structure of Galileo's critique of Aristotelian physics.

10. Trace in much greater detail the development of Newtonian ideas such as mass and inertia, spelling out their interrelations.

11. Analyze fully the explanatory coherence of relativity theory and quantum theory, with greater attention to the ways in which particular parts of those theories figure in particular explanations.

12. Analyze the conceptual and explanatory developments of additional cases in the social sciences, such as Keynesian economics and Chomskyan linguistics.

13. Analyze conceptual change and explanatory coherence in other important episodes in the history of science, such as the adoption of the atomic theory of matter and the wave theory of light in the early nineteenth century.

14. Develop more powerful computational models of abduction and conceptual combination.

If successful, such projects will further support the central claim of this book, that understanding of the development of science can be greatly enriched by looking at conceptual systems, conceptual change, and explanatory coherence.

10.4 SUMMARY

Scientific revolutions differ in the extent to which they have been data-driven, explanation-driven, and coherence-driven. They also differ in the kinds of conceptual change that they involved, although all major revolutions in the natural sciences include the most dramatic kinds of conceptual change:

branch jumping or tree switching. In every revolution, the new theory had greater explanatory coherence than the old theory, including greater explanatory breadth. But revolutions differ in the additional factors of explanatory coherence. The explanations employed by the revolutions all have a deductive strand, but they differ in the additional strands of explanation that were important. New theories usually supplant older ones, rejecting most of them, but the twentieth-century revolutions in physics are better described in terms of sublation, with new theories incorporating as well as rejecting Newtonian mechanics. From research to date, conceptual development in children appears to be less revolutionary than what occurs in science, since it does not involve branch jumping or tree switching. Whether explanatory coherence plays a role in theory development in children is an open question. There are many inviting projects for further investigation of conceptual and theoretical development in scientists and children.

References

Abelson, R., and Lalljee, M. (1988). Knowledge structures and causal explanation. In D. Hilton (Ed.), *Contemporary science and natural explanation.* New York: New York University Press, 175–203.

Achinstein, P. (1983). *The nature of explanation.* Oxford: Oxford University Press.

Ahn, W., and Brewer, W. (1991). Psychological studies of explanation-based learning. In G. DeJong (Ed.), *Studies in explanation-based learning.* Dordrecht: Kluwer, in press.

Anderson, A., and Belnap, N. (1975). *Entailment.* Princeton: Princeton University Press.

Anderson, J. R. (1980). *Cognitive psychology and its implications.* San Francisco: Freeman.

———. (1983). *The architecture of cognition.* Cambridge, MA: Harvard University Press.

Angell, J. (1908). *Psychology.* Fourth edition. New York: Holt.

Aristotle (1961). *Metaphysics.* Trans. J. Warrington. London: Dent.

———. (1984). *The complete works of Aristotle.* Ed. J. Barnes. Princeton: Princeton University Press.

Armstrong, S.; Gleitman, L.; and Gleitman, H. (1983). What some concepts might not be. *Cognition, 13,* 263–308.

Astington, J.; Harris, P.; and Olson, D. (1988). *Developing theories of mind.* Cambridge: Cambridge University Press.

Baars, B. (1986). *The cognitive revolution in psychology.* New York: Guilford.

Barrett, P.; Weinshank, D.; and Gottleber, T. (1981). *A concordance to Darwin's Origin of Species.* Ithaca: Cornell University Press.

Barwise, J., and Perry, J. (1983). *Situations and attitudes.* Cambridge, MA: MIT Press.

Bereiter, C., and Scardamalia, M. (1989). When weak explanations prevail. *Behavioral and Brain Sciences, 12,* 468–489.

Berkeley, G. (1962). *The principles of human knowledge.* London: Collins/Fontana. Originally published 1710.

Bloor, D. (1981). *Knowledge and social imagery.* London: Routledge and Kegan Paul.

Boas, M. (1962). *The scientific renaissance.* London: Fontana.

Bobrow, D. (Ed.), (1985). *Qualitative reasoning about physical systems.* Cambridge, MA: MIT Press.

BonJour, L. (1985). *The structure of empirical knowledge.* Cambridge, MA: Harvard University Press.

Bowler, P. (1984). *Evolution: The history of an idea.* Berkeley: University of California Press.

Brewer, W., and Samarapungavan, A. (1991). Children's theories versus scientific theories: Differences in reasoning or differences in knowledge? In R. Hoffman and D. Palermo (Eds.), *Cognition and the symbolic processes: Vol. 3: Applied and ecological perspectives.* Hillsdale, NJ: Erlbaum.

Bromberger, S. (1966). Why-questions. In R. Colodny (Ed.), *Mind and cosmos*. Pittsburgh: University of Pittsburgh Press, 86–111.

Brown, J. (Ed.) (1981). *Scientific rationality: The sociological turn*. Dordrecht: Reidel.

Bruner, J. S.; Goodnow, J. J.; and Austin, G.A. (1956). *A study of thinking*. New York: Wiley.

Brush, S. (1989). Prediction and theory evaluation: The case of light bending. *Science, 246,* 1124–1129.

Buchanan, B., and Shortliffe, E. (Eds.), (1984). *Rule-based expert systems*. Reading, MA: Addison Wesley.

Butterfield, H. (1965). *The origins of modern science*. Revised edition. New York: Free Press. Originally published 1957.

Bynum, W.; Browne, E.; and Porter, R. (1981). *Dictionary of the history of science*. Princeton: Princeton University Press.

Campbell, D. (1988). *Methodology and epistemology for social science: Selected papers*. Ed. E. Overman. Chicago: University of Chicago Press.

Campbell, N. (1957). *Foundations of science*. New York: Dover.

Carbonell, J. (Ed.), (1990). *Machine learning: Paradigms and methods*. Cambridge, MA: MIT Press.

Carey, S. (1985). *Conceptual change in childhood*. Cambridge, MA: MIT Press.

Carnap, R. (1950). *Logical foundations of probability*. Chicago: University of Chicago Press.

———. (1967). *The logical structure of the world*. Trans. R. George. Berkeley: University of California Press. German original published 1928.

Charniak, E., and McDermott, D. (1985). *Introduction to artificial intelligence*. Reading, MA: Addison-Wesley.

Chi, M. (1988). Children's lack of access and knowledge reorganization: An example from the concept of animism. In F. Weinert and M. Perlmutter (Eds.), *Memory development: Universal changes and individual differences*. Hillsdale, NJ: Erlbaum, 160–194.

———. (1991). Conceptual change within and across categories: Implications for learning and discovery in science. In R. Giere (Ed.), *Cognitive Models of Science, Minnesota Studies in the Philosophy of Science*, vol. 15. Minneapolis: University of Minnesota Press, in press.

Chomksy, N. (1959). Review of B. F. Skinner, *Verbal Behavior.Language*, *35*, 26–58.

———. (1988). *Language and problems of knowledge*. Cambridge, MA: MIT Press.

Chukovsky, K. (1984). *The art of translation*. Trans. L. Leighton. Knoxville: University of Tennessee Press.

Churchland, P. (1989). *A neurocomputational perspective*. Cambridge, MA: MIT Press.

Cohen, I. (1980) *The Newtonian revolution*. Cambridge: Cambridge University Press.

———. (1985). *Revolution in science*. Cambridge, MA: Harvard University Press.

Conant, J. (1964). *Harvard case histories in experimental science*, vol. 1. Cambridge, MA: Harvard University Press.

Copernicus (1959). *Three Copernican treatises*. Second edition. Trans. E. Rosen. New York: Dover.

———. (1978). *On the revolutions.* Trans. E. Rosen. Baltimore: Johns Hopkins University Press.

Cruse, D. (1986). *Lexical semantics.* Cambridge: Cambridge University Press.

Darden, L. (1991). *Theory change in science: Strategies from Mendelian genetics.* Oxford: Oxford University Press.

Darden, L., and Cain, J. (1989). Selection type theories. *Philosophy of Science, 56,* 106–129.

Darden, L., and Rada, R. (1988). Hypothesis formation using part-whole interrelations. In D. Hellman (Ed.), *Analogical reasoning.* Dordrecht: Reidel, 341–375.

Darwin, C. (1903). *More letters of Charles Darwin.* 2 vols. Ed. F. Darwin and A. Seward. London: John Murray.

———. (1958). Darwin's journal. In G. de Beer (Ed.), *Bulletin of the British Museum* (Natural History). Historical Series, London, vol. 2., no. 1.

———. (1962). *The origin of species.* Text of sixth edition. New York: Collier.

———. (1964). *On the origin of species.* Facsimile of first edition of 1859. Cambridge, MA: Harvard University Press.

———. (1981). *The descent of man, and selection in relation to sex.* Princeton: Princeton University Press. Originally published 1871.

———. (1987). *Charles Darwin's Notebooks, 1836–1844.* Ed. P. Barrett et al. Ithaca: Cornell University Press.

Davies, P. (1989). *The new physics.* Cambridge: Cambridge University Press.

Davies, P., and Brown, J. (1988). *Superstrings.* Cambridge: Cambridge University Press.

de Broglie, L. (1953). *The revolution in physics.* Trans. R. Niemeyer. New York: Noonday Press.

DeJong, G., and Mooney, R. (1986). Explanation-based learning: An alternative view. *Machine Learning, 1,* 145–176.

Descartes (1980). *Discourse on method and meditations on first philosophy.* Trans. D. Cress. Indianapolis: Hackett.

———. (1985). *The philosophical writings of Descartes.* Trans. J. Cottingham et al. Cambridge: Cambridge University Press.

Dietz, R. (1961). Continent and ocean basin evolution by spreading of the sea floor. *Nature, 190,* 854–857.

diSessa, A. (1988). Knowledge in pieces. In G. Forman and P. Pufall (Eds.), *Constructivism in the computer age.* Hillsdale, NJ: Erlbaum, 49–70.

Dolnick, E. (1989). Panda paradox. *Discover Magazine,* September, 71–72.

Donovan, A. (1988). *The chemical revolution: essays in reinterpretation. Osiris,* second series, vol. 4.

Dretske, F. (1981). *Knowledge and the flow of information.* Cambridge, MA: MIT Press.

Duhem, P. (1954). *The aim and structure of physical theory.* Trans. P. Wiener. Princeton: Princeton University Press. Originally published 1914.

Einhorn, H., and Hogarth, R. (1986). Judging probable cause. *Psychological Bulletin, 99,* 3–19.

Einstein, A. (1949). Autobiographical notes. In P. Schilpp (Ed.), *Albert Einstein: Philosopher-scientist.* La Salle, IL: Open Court, 2–93.

———. (1952). *The principle of relativity.* New York: Dover.

Einstein, A. (1961). *Relativity: The special and the general theory.* New York: Crown.

———., and Infeld, L. (1938). *The evolution of physics.* New York: Simon and Schuster.

Eldredge, N. (1985). *Time frames: The rethinking of Darwinian evolution and the theory of punctuated equilibrium.* New York: Simon and Schuster.

Falkenhainer, B.; Forbus, K.; and Gentner, D. (1989). The structure-mapping engine: Algorithms and examples. *Artificial Intelligence, 41*, 1–63.

Feldman, J., and Ballard, D. (1982). Connectionist models and their properties. *Cognitive Science 6*, 205–254.

Fellbaum, C., and Miller, G. (1990). Folk psychology or semantic entailment? Comment on Rips and Conrad. *Psychological Review, 97*, 565–570.

Feyerabend, P. (1965). Problems of empiricism. In R. Colodny (Ed.), *Beyond the edge of certainty.* Pittsburgh: University of Pittsburgh Press, 145–260.

———. (1975). *Against method.* London: New Left Books.

Feynman, R.; Leighton, R.; and Sands, M. (1963–1964). *The Feynman lectures on physics.* 2 vols. Reading, MA: Addison-Wesley.

Fiske, S., and Taylor, S. (1984). *Social cognition.* New York: Random House.

Fodor, J. (1975). *The language of thought.* New York: Crowell.

Fodor, J.; Garrett, M.; Walker, E.; and Parkes, C. (1980). Against definitions. *Cognition, 8*, 1–105.

Forbus, K. (1985). Qualitative process theory. In D. Bobrow (Ed.), *Qualitative reasoning about physical systems.* Cambridge, MA: MIT Press, 85–168.

Foster, M., and Martin, M. (Eds.), (1966). *Probability, confirmation, and simplicity.* New York: Odyssey Press.

Frankel, H. (1979). The career of continental drift theory: An application of Imre Lakatos' analysis of scientific growth to the rise of drift theory. *Studies in the History and Philosophy of Science, 10,* 21–66.

Freedman, E. (1991). Understanding scientific controversies from a computational perspective: The case of latent learning. In R. Giere (Ed.), *Cognitive Models of Science, Minnesota Studies in the Philosophy of Science*, vol. 15. Minneapolis: University of Minnesota Press.

Frege, G. (1968). *The foundations of arithmetic.* Trans. J. Austin. Oxford: Basil Blackwell. German original published 1884.

———. (1970). *Translations from the philosophical writings of Gottlob Frege.* Ed. P. Geach and M. Black. Oxford: Basil Blackwell.

Friedman, M. (1974). Explanation and scientific understanding. *Journal of Philosophy, 71*, 5–19.

———. (1983). *Foundations of space-time theories.* Princeton: Princeton University Press.

Galileo (1957). *Discoveries and opinions of Galileo.* Trans. S. Drake. New York: Doubleday.

Gamov, G. (1985). *Thirty years that shook physics.* New York: Dover.

Gärdenfors, P. (1988). *Knowledge in flux.* Cambridge, MA: MIT Press/Bradford Books.

Gardner, H. (1985). *The mind's new science.* New York: Basic Books.

Genesereth, M., and Nilsson, N. (1987). *Logical foundations of artificial intelligence.* Los Altos: Morgan Kaufmann.

Gick, M. L., and Holyoak, K. J. (1983). Schema induction and analogical transfer. *Cognitive Psychology, 15*, 1–38.

Giere, R. (1988). *Explaining science: A cognitive approach.* Chicago: University of Chicago Press.

Gilbert, F. (1973). Revolution. In P. Wiener (Ed.), *Dictionary of the history of ideas*, vol. 4. New York: Scribners, 152–167.

Glen, W. (1975). *Continental drift and plate tectonics.* Columbus, OH: Charles E. Merrill Publishing Co.

Goldman, A. (1986). *Epistemology and cognition.* Cambridge, MA: Harvard University Press.

Grosjean, F. (1982). *Life with two languages.* Cambridge, MA: Harvard University Press.

Guerlac, H. (1961). *Lavoisier—The crucial year.* Ithaca: Cornell University Press.

Guillemin, V. (1968). *The story of quantum mechanics.* New York: Scribner's.

Hacking, I. (1975). *Why does language matter to philosophy?* Cambridge: Cambridge University Press.

Hallam, A. (1973). *A revolution in the earth sciences.* Oxford: Clarendon Press.

Hammond, K. (1989). *Case-based planning: Viewing planning as a memory task.* New York: Academic Press.

Hanson, N. R. (1958). *Patterns of discovery.* Cambridge: Cambridge University Press.

Hanson, S., and Burr, D. (1990). What connectionist models learn: Learning and representation in connectionist networks. *Behavioral and Brain Sciences, 13*, 471–518.

Harman, G. (1973). *Thought.* Princeton: Princeton University Press.

———. (1986). *Change in view: Principles of reasoning.* Cambridge, MA: MIT Press/Bradford Books.

———. (1987). (Nonsolopsistic) conceptual role semantics. In E. LePore (Ed.), *Semantics of natural language.* New York: Academic Press, 55–81.

Harman, G.; Ranney, M.; Salem, K.; Doring, F.; Epstein, J.; and Jaworksa, A. (1988). A theory of simplicity. In *Proceedings of the Tenth Annual Conference of the Cognitive Science Society.* Hillsdale, NJ: Erlbaum, 111–117.

Harvey, W. (1962). *On the motion of the heart and blood in animals.* Trans. A. Bowie. Chicago: Henry Regnery. Originally published 1628.

Hausman, D. (1982). Causal and explanatory asymmetry. In P. Asquith and T. Nickles (Eds.), *PSA 1982*, vol. 1. East Lansing, MI: Philosophy of Science Association, 43–54.

Hedges, L. (1987). How hard is hard science, how soft is soft science? *American Psychologist, 42*, 443–455.

Hegel, G. (1967). *The phenomenology of mind.* Trans. J. Baillie. New York: Harper and Row. Originally published 1807.

Heltai, P. (1988). Hierarchical lexical relations in English and Hungarian. Unpublished manuscript, Budapest.

Hempel, C. (1952). *Fundamentals of concept formation in natural science.* Chicago: University of Chicago Press.

———. (1965). *Aspects of scientific explanation.* New York: Free Press.

Hess, H. (1962). History of ocean basins. In A. Engel, H. James, and B. Leonard, (Eds.), *Petrologic studies: A volume to honor A. F. Buddington.* New York: Geological Society of America, 599–620.

Hesse, M. (1966). *Models and analogies in science.* Notre Dame, IN: Notre Dame University Press.

Hinton, G. (1988). Representing part-whole hierarchies in connectionist networks. In *Proceedings of the Tenth Annual Conference of the Cognitive Science Society.* Hillsdale, NJ: Erlbaum, 48–54.

Hinton, G., and Anderson, J. (Eds.), (1981). *Parallel models of associative memory.* Hillsdale, NJ: Erlbaum.

Hobbs, J. (1989). Are explanatory coherence and a connectionist model necessary? *Behavioral and Brain Sciences, 12*, 476–477.

Hobbs, J.; Stickel, M.; Appelt, D.; and Martin, P. (1990). Interpretation as abduction. Technical Note 499, SRI International.

Holland, J. (1986). Escaping brittleness: The possibilities of general purpose machine learning algorithms applied to parallel rule-based systems. In R. S. Michalski, J. G. Carbonell, and T. M. Mitchell (Eds.), *Machine learning: An artificial intelligence approach*, vol. 2. Los Altos: Kaufmann, 593–623.

Holland, J.; Holyoak, K.; Nisbett, R.; and Thagard, P. (1986). *Induction: Processes of inference, learning, and discovery.* Cambridge, MA: MIT Press/Bradford Books.

Holmes, F. (1985). *Lavoisier and the chemistry of life.* Madison: University of Wisconsin Press

Holton, G. (1973). *Thematic origins of scientific thought: Kepler to Einstein.* Cambridge, MA: Harvard University Press.

Holton, G., and Brush, S. (1985). *Introduction to concepts and theories in physical science.* Second edition. Princeton: Princeton University Press.

Holyoak, K.; Koh, K.; and Nisbett, R. (1989). A theory of conditioning: Inductive learning within rule-based default hierarchies. *Psychological Review, 96*, 315–340.

Holyoak, K., and Thagard, P. (1989). Analogical mapping by constraint satisfaction. *Cognitive Science, 13*, 295–355.

Hopfield, J., and Tank, D. (1985). Neural computation of decisions in optimization problems. *Biological Cybernetics, 52,* 141–152.

Horwich, P. (1982). *Probability and evidence.* Cambridge: Cambridge University Press.

Hull, C. (1943). *Principles of behavior.* New York: Appleton-Century-Crofts.

Hull, D. (Ed.), (1973). *Darwin and his critics.* Chicago: University of Chicago Press.

———. (1989). *Science as a process.* Chicago: University of Chicago Press.

Hume, D. (1888). *A treatise of human nature.* Ed. L. A. Selby-Bigge. London: Oxford University Press. Originally published 1739.

Jacobs, J.; Russell, R.; and Wilson, J. (1959). *Physics and geology.* New York: McGraw-Hill.

James, W. (1961). *Psychology: The briefer course.* New York: Harper. Originally published 1892.

———. (1982). *The varieties of religious experience.* Harmondsworth: Penguin. Originally published 1902.

Jammer, M. (1966). *The conceptual development of quantum mechanics.* New York: McGraw-Hill.

Josephson, J.; Chandrasekaran, B.; Smith, J.; and Tanner, M. (1987). A mechanism for forming composite explanatory hypotheses. *IEEE Transactions on Systems, Man, and Cybernetics*, 17, 445–454.

Kant, I. (1965). *Critique of pure reason*. Second edition. Trans. N. Kemp Smith. London: MacMillan. Originally published 1787.

Keil, F. (1989). *Concepts, kinds, and cognitive development*. Cambridge, MA: MIT Press/Bradford Books.

Kirwan, R. (1968). *An essay on phlogiston and the constitution of acids*. New impression of second English edition of 1789. London: Cass.

Kitcher, P. (1981). Explanatory unification. *Philosophy of Science*, *48*, 507–531.

———. (1988). The child as parent of the scientist. *Mind & Language, 3,* 217–228.

Kitcher, P., and Salmon, W. (1989). *Scientific explanation*. Minneapolis : University of Minnesota Press.

Knickerbocker, W. (1962). *Classics of modern science*. Boston: Beacon.

Kuhn, D., et al. (1988). *The development of scientific thinking skills*. Orlando, FL : Academic Press, 1988.

Kuhn, T. (1957). *The Copernican revolution*. Cambridge: Harvard University Press.

———. (1970). *Structure of scientific revolutions*. Second edition. Chicago: University of Chicago Press. Originally published 1962.

———. (1977). *The essential tension*. Chicago: University of Chicago Press

———. (1978). *Black-body theory and the quantum discontinuity*. Oxford: Oxford University Press.

———. (1983). Commensurability, comparability, communicability. In P. Asquith and T. Nickles (Eds.), *PSA 1982*, vol. 2. East Lansing, MI: Philosophy of Science Association, 669–688.

———. (1987). What are scientific revolutions? In L. Kruger et al. (Eds.), *The probabilistic revolution*, vol. 1. Cambridge, MA: MIT Press/Bradford Books, 7–22.

Kunda, Z. (1987). Motivation and inference: Self-serving generation and evaluation of causal theories. *Journal of Personality and Social Psychology, 53,* 636–647.

———. (1990). The case for motivated inference. *Psychological Bulletin, 108,* 480–498.

Kunda, Z.; Miller, D.; and Claire, T. (1990). Combining social concepts: The role of causal reasoning. *Cognitive Science, 14,* 551–577.

Laird, J.; Rosenbloom, P.; and Newell, A. (1986). Chunking in SOAR: The anatomy of a general learning mechanism. *Machine Learning, 1,* 11–46.

Lakatos, I. (1970). Falsification and the methodology of scientific research programs. In I. Lakatos and A. Musgrave (Eds.), *Criticism and the growth of knowledge*. Cambridge: Cambridge University Press, 91–195.

Lakoff, G. (1987). *Women, fire, and dangerous things*. Chicago: University of Chicago Press.

Langley, P.; Simon, H.; Bradshaw, G.; and Zytkow, J. (1987). *Scientific discovery*. Cambridge, MA: MIT Press/Bradford Books.

Larkin, J.; Reif, F.; Carbonell, J.; and Gugliotta, A. (1988). A flexible expert reasoner with multi-domain inferencing. *Cognitive Science, 12,* 101–138.

Larson, M. (1984). *Meaning-based translation: A guide to cross-language equivalence*. Lanham, MD: University Press of America.

Laudan, L. (1977). *Progress and its problems*. Berkeley: University of California Press.

Lavoisier, A. (1789). *Traité élémentaire de chimie*. Paris: Cuchet. English translation, 1790: *Elements of chemistry*. Trans. R. Kerr. Edinburgh: W. Creech.

Lavoisier, A. (1862). *Oeuvres*. 6 vols. Paris: Imprimerie Impériale.

———. (1970). *Essays physical and chemical*. Trans. T. Henry, 1776. Second English edition. London: Cass. Originally published 1774.

Leibniz, G. (1951). *Selections*. Ed. P. Wiener. New York: Scribner's.

Leicester, H., and Krickstein, H. (1952). *A source book in chemistry*. New York: McGraw-Hill.

Lewis, D. (1990). *Parts of classes*. Oxford: Basil Blackwell.

Lloyd, E. (1983). The nature of Darwin's support for the theory of natural selection. *Philosophy of Science*, *50*, 112–129.

Locke, J. (1961). *An essay concerning human understanding*. London: Dent. Originally published 1690.

Markman, E. (1989). *Categorization and naming in children*. Cambridge, MA: MIT Press/Bradford Books.

Mayr, E. (1982). *The growth of biological thought*. Cambridge, MA: Harvard University Press.

McCann, H. (1978). *Chemistry transformed: the paradigmatic shift from phlogiston to oxygen*. Norwood, NJ: Ablex.

McCauley, R. (1986). Intertheoretic relations and the future of psychology. *Philosophy of Science*, *53*, 179–199.

McCloskey, M. (1983). Intuitive physics. *Scientific American*, *24*, 122–130.

———. (1983). Naive theories of motion. In D. Gentner and A. Stevens (Eds.), *Mental Models*. Hillsdale, NJ: Erlbaum, 299–324.

Medin, D. (1989). Concepts and conceptual structure. *American Psychologist*, *44*, 1469–1481.

Medin, D., and Shoben, E. (1988). Context and structure in conceptual combination. *Cognitive Psychology*, 20, 158–190.

Menard, H. (1986). *The ocean of truth: A personal history of global tectonics*. Princeton: Princeton University Press.

Michalski, R.; Carbonell, J.; and Mitchell, T. (Eds.), (1983). *Machine learning: An artificial intelligence approach*. Palo Alto: Tioga.

———. (1986). *Machine Learning*, vol. 2. Los Altos: Morgan Kaufmann.

Miller, G. (1956). The magical number seven, plus or minus two: Some limits on our capacity for processing information. *Psychological Review*, *63*, 81–97.

———. (1965). Some preliminaries to psycholinguistics. *American Psychologist*, *20*, 15–20.

Miller, G.; Beckwith, R.; Fellbaum, C.; Gross, G.; and Miller, K. (1990). Introduction to WordNet: An on-line lexical database. *International Journal of Lexicography*, *3*, 235–244.

Miller, G.; Galanter, E.; and Pribram, K. (1960). *Plans and the structure of behavior*. New York: Holt, Rinehart, and Winston.

Miller, G., and Johnson-Laird, P. (1976). *Language and perception*. Cambridge, MA: Harvard University Press.

Miller, L., and Read, S. (1991). On the coherence of mental models of persons and relationships. In G. Fletcher and F. Fincham (Eds.), *Cognition in close relationships*. Hillsdale, NJ: Erlbaum, 69–99.

Miller, R. (1987). *Fact and method*. Princeton: Princeton University Press.

Minsky, M. (1975). A framework for representing knowledge. In P. H. Winston (Ed.), *The psychology of computer vision*. New York: McGraw-Hill, 211–277.

Mitchell, T.; Keller, R.; and Kedar-Cabelli, S. (1986). Explanation-based generalization: A unifying view. *Machine Learning, 1,* 47–80.

Mooney, R., and DeJong, G. (1985). Learning schemata for natural language processing. In *Proceedings of the Ninth International Joint Conference on Artificial Intelligence.* San Mateo: Morgan Kaufmann, 681–687.

Morgan, W. J. (1968). Rises, trenches, great faults, and crustal blocks. *Journal of Geophysics Research, 73,* 1959–1982.

Murphy, G., and Medin, D. (1985). The role of theories in conceptual coherence. *Psychological Review*, *92*, 289–316.

Neisser, U. (1967). *Cognitive psychology*. New York: Appleton-Century-Crofts.

———. (1976). *Cognition and reality*. San Francisco: W. H. Freeman.

Nelson, G., Thagard, P., and Hardy, S. (1991). Integrating analogies with rules and explanations. In J. Barnden and K. Holyoak (Eds.), *Advances in Connectionism*, vol. 2, *Analogical Connections*. Norwood, NJ: Ablex.

Nersessian, N. (1989). Conceptual change in science and in science education. *Synthese, 80,* 163–183.

Newell, A. (1990). *Unified theories of cognition*. Cambridge, MA: Harvard University Press.

Newell, A.; Shaw, C.; and Simon, H. (1958). Elements of a theory of human problem solving. *Psychological Review*, *65*, 151–166.

Newell, A., and Simon, H. A. (1972). *Human problem solving*. Englewood Cliffs, NJ: Prentice-Hall.

Newton, I. (1934). *Mathematical principles of natural philosophy*. Trans. A. Motte and F. Cajori. Berkeley: University of California Press. Originally published 1726.

Nisbett, R. E., and Ross, L. (1980). *Human inference: Strategies and shortcomings of social judgment*. Englewood Cliffs, NJ: Prentice-Hall.

Nisbett, R. E., and Wilson, T. D. (1977). Telling more than we can know: Verbal reports on mental processes. *Psychological Review*, *84*, 231–259.

Nowak, G., and Thagard, P. (1991a). Copernicus, Ptolemy, and explanatory coherence. In R. Giere (Ed.), *Cognitive Models of Science, Minnesota Studies in the Philosophy of Science*, vol. 15. Minneapolis: University of Minnesota Press, in press.

———. (1991b). Newton, Descartes, and explanatory coherence. In R. Duschl and R. Hamilton (Eds)., *Philosophy of science, cognitive psychology and educational theory and practice*. Albany: SUNY Press, in press.

Oldroyd, D. (1980). *Darwinian impacts*. Milton Keynes: Open University Press.

O'Rorke, P.; Morris, S.; and Schulenburg, D. (1990). Theory formation by abduction: A case study based on the chemical revolution. In J. Shrager and P. Langley (Eds.), *Computational models of discovery and theory formation*. San Mateo: Morgan Kaufmann, 197–224.

Pais, A. (1982). *"Subtle is the Lord. . .": The science and life of Albert Einstein.* Oxford: Oxford University Press.

Paley, W. (1963). *Natural Theology*. Ed. F. Ferré. Indianapolis: Bobbs-Merrill. Selections from work published in 1802.

Partington, J. (1961). *A history of chemistry*. 4 vols. London: Macmillan.

Pearl, J. (1988). *Probabilistic reasoning in intelligent systems*. San Mateo: Morgan Kaufmann.

Peirce, C. (1931–1958). *Collected papers*. 8 vols. Ed. C. Hartshorne, P. Weiss, and A. Burks. Cambridge, MA: Harvard University Press.

Peng, Y., and Reggia, J. (1990). *Abductive inference models for diagnostic problem solving*. New York: Springer-Verlag.

Pennington, N., and Hastie, R. (1986). Evidence evaluation in complex decision making. *Journal of Personality and Social Psychology, 51*, 242–258.

———. (1987). Explanation-based decision making. In *Proceedings of the Ninth Annual Conference of the Cognitive Science Society*. Hillsdale, NJ: Erlbaum, 682–690.

Perrin, C. (1988). The chemical revolution: Shifts in guiding assumptions. In A. Donovan, L. Laudan, and R. Laudan (Eds.), *Scrutinizing science: Empirical studies of scientific change*. Dordrecht: Kluwer, 105–124.

Plato (1961). *The Collected Dialogues*. Ed. E. Hamilton and H. Cairns. Princeton: Princeton University Press.

Pollock, J. (1989). *How to build a person: A prolegomenon*. Cambridge, MA: MIT Press/Bradford Books.

Pople, H. (1973). On the mechanization of abductive logic. In *Proceedings of the Third International Joint Conference on Artificial Intelligence*. San Mateo: Morgan Kaufmann, vol. 2, 147–152.

———. (1977). The formation of composite hypotheses in diagnostic problem solving. In *Proceedings of the Fifth International Joint Conference on Artificial Intelligence*. San Mateo: Morgan Kaufmann, 1030–1037.

Popper, K. (1959). *The logic of scientific discovery*. London: Hutchinson.

Priestley, J. (1929). *Considerations on the doctrine of phlogiston, and the decomposition of water*. Princeton: Princeton University Press. Originally published 1796.

Ptolemy (1984). *Ptolemy's Almagest*. Trans. G. Toomer. London: Duckworth.

Putnam, H. (1975). *Mind, language, and reality*. Cambridge: Cambridge University Press.

Quillian, M. (1968). Semantic memory. In M. Minsky (Ed.), *Semantic information processing*. Cambridge, MA: MIT Press, 227–270.

Quine, W.V.O. (1960). *Word and object*. Cambridge, MA: MIT Press.

———. (1963). *From a logical point of view*. Second edition. New York: Harper Torchbooks.

Railton, P. (1978). A deductive-nomological model of probabilistic explanation. *Philosophy of Science, 45*, 206–226.

Ranney, M. (1991). Explorations in explanatory coherence. In E. Bar-On, B. Eylon, and Z. Schertz (Eds.), *Designing intelligent learning environments: From cognitive analysis to computer implementation*. Norwood, NJ: Ablex, in press.

Ranney, M., and Thagard, P. (1988). Explanatory coherence and belief revision in naive physics. In *Proceedings of the Tenth Annual Conference of the Cognitive Science Society*. Hillsdale, NJ: Erlbaum, 426–432.

Recker, D. (1987). Causal efficacy: The structure of Darwin's argument strategy in the *Origin of Species*. *Philosophy of Science, 54*, 147–175.

Rescorla, R. (1988). Pavlovian conditioning: It's not what you think. *American Psychologist, 43,* 151–160.

Ridley, M. (1986). *Evolution and classification.* London: Longman.

Rips, L. (1990). Intuitive psychologists: Conceptions of mental activities and their parts. In J-C. Smith (Ed.), *Essays on the historical foundations of cognitive science.* Dordrecht: Kluwer, 267–292.

Rips, L., and Conrad, F. (1989). Folk psychology of mental activities. *Psychological Review, 96,* 187–207.

Rosch, E.; Mervis, C. B.; Gray, W.; Johnson, D.; and BoyesBraem, P. (1976). Basic objects in natural categories. *Cognitive Psychology, 7,* 573–605.

Rose, D., and Langley, P. (1986). Chemical discovery as belief revision. *Machine Learning, 1,* 423–452.

Rumelhart, D.; McClelland, J.; and the PDP Research Group. (1986). *Parallel distributed processing: Explorations in the microstructure of cognition,* 2 vols. Cambridge, MA: MIT Press/Bradford Books.

Rumelhart, D.; Smolensky, P.; Hinton, G.; and McClelland, J. (1986). Schemata and sequential thought processes in PDP models. In J. McClelland and D. Rumelhart (Eds.), *Parallel distributed processing: Explorations in the microstructure of cognition,* vol. 2. Cambridge, MA: MIT Press/Bradford Books, 7–57.

Ruse, M. (1979). *The Darwinian revolution.* Chicago: University of Chicago Press.

Salmon, W. (1966). *The foundations of scientific inference.* Pittsburgh: University of Pittsburgh Press.

———. (1970). Statistical explanation. In R. Colodny (Ed.), *The nature and function of scientific theories.* Pittsburgh: University of Pittsburgh Press, 173–231.

———. (1984). *Scientific explanation and the causal structure of the world.* Princeton: Princeton University Press.

Salzman, L. (1953). The psychology of religious and ideological conversion. *Psychiatry, 16,* 177–187.

Samarapungavan, A. (1990). Children's metajudgments in theory choice tasks: An investigation of scientific rationality in childhood. Unpublished dissertation, University of Illinois at Urbana-Champaign.

Sambursky, S. (1956). *The physical world of the Greeks.* London: Routledge and Kegan Paul.

Schank, R. (1986). *Explanation patterns.* Hillsdale, NJ: Erlbaum.

Schank, R., and Abelson, R. P. (1977). *Scripts, plans, goals, and understanding: An inquiry into human knowledge structures.* Hillsdale, NJ: Erlbaum.

Shannon, C., and Weaver, W. (1949). *The mathematical theory of communication.* Urbana: University of Illinois Press.

Shapere, D. (1974). *Galileo.* Chicago: University of Chicago Press.

Simon, H. (1945). *Administrative behavior.* New York: Free Press.

Skinner, B. F. (1953). *Science and human behavior.* New York: Free Press.

———. (1976). *About behaviorism.* New York: Vintage.

Skocpol, T. (1979). *States and social revolutions.* Cambridge: Cambridge University Press.

Slezak, P. (1989). Scientific discovery by computer as empirical refutation of the strong programme. *Social Studies of Science, 19,* 563–600.

Smart, J. J. C. (1989). Explanation. Unpublished address to the Royal Institute of Philosophy.

Smith, E. (1989). Concepts and induction. In M. Posner (Ed.), *Foundations of cognitive science*. Cambridge, MA: MIT Press, 501–526.

Smith, E.; Osherson, D.; Rips, L.; and Keane, M. (1988). Combining prototypes: A selective modification model. *Cognitive Science, 12*, 485–527.

Smith, L. (1986). *Behaviorism and logical positivism*. Stanford: Stanford University Press.

Stahl, G. (1730). *Philosophical principles of universal chemistry.* Trans. P. Shaw from *Fundamenta Chymiae*, 1723. London: John Osborn and Peter Longman.

Taylor, C. (1975). *Hegel*. Cambridge: Cambridge University Press.

Thagard, P. (1978). The best explanation: Criteria for theory choice. *Journal of Philosophy*, *75*, 76–92.

———. (1982). Hegel, science, and set theory. *Erkenntnis*, *18*, 397–410.

———. (1984). Frames, knowledge, and inference. *Synthese*, *61*, 233–259.

———. (1988). *Computational philosophy of science*. Cambridge, MA: MIT Press/Bradford Books.

———. (1989). Explanatory coherence. *Behavioral and Brain Sciences*, *12*, 435–467.

———. (1991a). Adversarial problem solving: Modeling an opponent using explanatory coherence. Unpublished manuscript, Princeton University.

———. (1991b) The dinosaur debate: Explanatory coherence and the problem of competing hypotheses. In J. Pollock and R. Cummins (Eds.), *Philosophy and AI: Essays at the interface*. Cambridge, MA: MIT Press/Bradford Books, in press.

———. (1991c). Probabilistic networks and explanatory coherence. To appear in a volume on abductive inference edited by P. O'Rorke and J. Josephson.

Thagard, P.; Cohen, D.; and Holyoak, K. (1989). Chemical analogies: Two kinds of explanation. In *Proceedings of the Eleventh International Joint Conference on Artificial Intelligence*. San Mateo: Morgan Kaufmann, 819–824.

Thagard, P. and Holyoak, K. (1985). Discovering the wave theory of sound: Induction in the context of problem solving. In *Proceedings of the Ninth International Joint Conference on Artificial Intelligence*. Los Altos: Morgan Kaufmann, 610–612.

Thagard, P.; Holyoak, K.; Nelson, G.; and Gochfeld, D. (1990). Analog retrieval by constraint satisfaction. *Artificial Intelligence, 46*, 259–310.

Thorndike, E. L. (1949). *Selected writings from a connectionist's psychology*. New York: Appleton-Century-Crofts.

Tolman, E. C. (1932). *Purposive behavior in animals and men*. New York: Century Co.

Toulmin, S. (1972). *Human understanding*. Princeton: Princeton University Press.

Touretzky, D., and Hinton, G. (1988). A distributed production system. *Cognitive Science, 12*, 423–466.

Tversky, B. (1989). Parts, partonomies, and taxonomies. *Developmental Psychology, 25*, 983–995.

van der Gracht, W., et al. (1928). *Theory of continental drift*. Tulsa, OK: American Association of Petroleum Geologists.

van Fraassen, B. (1980). *The scientific image*. Oxford: Clarendon Press.

———. (1989). *Laws and symmetries.* Oxford: Clarendon Press.

Velikovsky, I. (1965). *Worlds in collision*. New York: Dell.

Vine, F., and Matthews, D. (1963). Magnetic anomalies over ocean ridges. *Nature, 199*, 947–949.

Vosniadou, S., and Brewer, W. (1987). Theories of knowledge restructuring in development. *Review of Educational Research, 57*, 51–67.

———. (1990). Mental models of the earth: A study of conceptual change in childhood. Unpublished manuscript, University of Illinois.

Walden, P. (1929). *Salts, acids and bases*. New York: McGraw-Hill.

Watson, J. (1913). Psychology as the behaviorist views it. *Psychological Review, 20*, 158–177.

———. (1917). An attempted formulation of the science of behavior psychology. *Psychological Review, 24*, 329–352.

———. (1929). *Psychology from the standpoint of a behaviorist.* Third edition. Philadelphia: Lippincott. Originally published 1919.

———. (1959). *Behaviorism.* Second impression of 1930 edition. Chicago: University of Chicago Press. Originally published 1924.

Wegener, A. (1966). *The origin of continents and oceans.* Trans. J. Biram, from fourth revised German edition. New York: Dover. Originally published 1929. First edition 1915.

Weitz, M. (1988). *Theories of concepts*. London: Routledge.

Wellman, H. (1990). *The child's theory of mind.* Cambridge, MA: MIT Press/Bradford Books.

Whorf, B. (1956). *Language, thought, and reality.* Cambridge, MA: MIT Press.

Will, C. (1986). *Was Einstein right? Putting general relativity to the test*. New York: Basic Books.

Wilson, J. Tuzo (1954). The development and structure of the crust. In G. Kuiper (Ed.), *The earth as a planet.* Chicago: University of Chicago Press, 138–214.

———. (1963). Continental drift. *Scientific American, 208*, 86–100.

———. (1965). A new class of faults and their bearing on continental drift. *Nature, 207*, 343–347.

Winston, M.; Chaffin, R.; and Herrmann, D. (1987). A taxonomy of part-whole relations. *Cognitive Science, 11*, 417–444.

Wittgenstein, L. (1953). *Philosophical investigations*. Trans. G. E. M. Anscombe. Oxford: Basil Blackwell.

Index